HOLD PARAMOUNT
The Engineer's Responsibility to Society

HOLD PARAMOUNT
The Engineer's Responsibility to Society

Alastair S. Gunn
University of Waikato Hamilton, New Zealand

P. Aarne Vesilind
Bucknell University
Lewisburg, Pennsylvania

Australia • Canada • Mexico • Singapore • Spain • United Kingdom • United States

Publisher: *Bill Stenquist*
Editorial Coordinator: *Valerie Boyajian*
Project Manager, Editorial Production: *Mary Vezilich*
Print/Media Buyer: *Vena Dyer*
Production Service: *Matrix Productions Inc.*

Cover and Text Designer: *Roy R. Neuhaus*
Copy Editor: *Jan McDearmon*
Printing and Binding: *Transcontinental–Louisville*
Compositor: *Nighthawk*

COPYRIGHT © 2003 by Brooks/Cole, a division of Thomson Learning, Inc. Thomson Learning™ is a trademark used herein under license.

ALL RIGHTS RESERVED. No part of this work covered by the copyright hereon may be reproduced or used in any form or by any means—graphic, electronic, or mechanical, including but not limited to photocopying, recording, taping, Web distribution, information networks, or information storage and retrieval systems—without the written permission of the publisher.

All products used herein are used for identification purpose only and may be trademarks or registered trademarks of their respective owners.

Printed in Canada
1 2 3 4 5 6 7 06 05 04 03 02

For more information about our products, contact us at:
Thomson Learning Academic Resource Center
1-800-423-0563
For permission to use material from this text, contact us by:
Phone: 1-800-730-2214
Fax: 1-800-730-2215
Web: http://www.thomsonrights.com

Library of Congress Control Number: 2002027742

ISBN 0-534-39258-X

Brooks/Cole–Thomson Learning
511 Forest Lodge Road
Pacific Grove, CA 93950
USA

Asia
Thomson Learning
5 Shenton Way #01-01
UIC Building
Singapore 068808

Australia
Nelson Thomson Learning
102 Dodds Street
South Melbourne, Victoria 3205
Australia

Canada
Nelson Thomson Learning
1120 Birchmount Road
Toronto, Ontario M1K 5G4
Canada

Europe/Middle East/Africa
Thomson Learning
High Holborn House
50/51 Bedford Row
London WC1R 4LR
United Kingdom

*We dedicated our first collaboration to our ancestors,
the second to our wives,
and this book is for our children:
James and Laura Gunn, and
Pamela, Steve, and Lauren Vesilind*

About the Authors

Alastair S. Gunn

The lead author is an Associate Professor and Chair of the Department of Philosophy, University of Waikato, Hamilton, New Zealand, where he teaches and researches several areas of applied ethics. He is also Associate Dean for e-learning in the faculty of arts and social sciences. He has taught at several universities in the U.S. and Southeast Asia and carries out ethics consultancies in New Zealand and internationally. He is a member of several ethics boards and has represented New Zealand at a number of international workshops and meetings.

P. Aarne Vesilind

The co-author is the R. L. Rooke Professor of Engineering at Bucknell University in Lewisburg, Pennsylvania. He teaches courses in professional ethics and the impact of technology on society, as well as more technical courses in water and wastewater treatment in the College of Engineering. He served on the faculty of Duke University for nearly 30 years where he directed the Center for Applied Ethics. He has lectured at many universities and has been a Fullbright Fellow at the University of Waikato.

Contents

Preface xi
About this Book: A Note to the Instructor xiii

1 Doing the right thing 1

1-1 Deception I 2
1-2 Keeping Promises 3
1-3 Doing the Right Thing 4
1-4 Theft of Music 6
1-5 Obligation to Strangers 7
1-6 Moral Rules 8
    Discussion Questions 8

2 The engineering profession 10

2-1 On Being a Professional 11
2-2 Technical Expertise and Ethical Obligations 12
2-3 Organization of Professional Engineering 12
2-4 Can We Afford to Be Ethical? 14
2-5 Engineering Codes of Ethics 14
2-6 Can a Person Stop Being an Engineer? 15
2-7 Codes of Ethics and the Environment 16
2-8 Ethically Right for Me? I 18
2-9 Ethical Theories as Decision-Making Models 19
    Discussion Questions 22

3 Enhance human welfare 25

3-1 Moral Responsibilities of Engineers 27
3-2 Engineering and Terrorism 28
3-3 Engineers as Intelligent Robots 29
    Discussion Questions 29

4 Hold paramount 31

- 4-1 Why Can't Ethicists Be as Efficient as Engineers? 32
- 4-2 Medical Ethics 33
- 4-3 Legal Ethics 35
- 4-4 Jokes about Engineers 36
- 4-5 Engineers Working Together 37
- 4-6 A Technical Challenge 38
- 4-7 Engineering Qualifications 41
- 4-8 Factors of Safety 41
- 4-9 Engineering Triumphs 43
- 4-10 Engineering Failures 44
- 4-11 Engineers as Managers 46
- 4-12 Acceptable Risk 48
- 4-13 Decision Making: Technical and Ethical Aspects 51
- 4-14 Consulting with Colleagues 53
- Discussion Questions 54

5 Safety of the public 56

- 5-1 The Moral Status of Animals 56
- 5-2 Ethical and Legal Obligations 58
- 5-3 Ethical Dilemmas I 59
- 5-4 Calculating the Value of Life 59
- 5-5 Fix Up Your Organization Ethically? 60
- 5-6 Whistle-Blowing I 62
- 5-7 Whistle-Blowing II 63
- 5-8 Disaster in Kansas City 66
- 5-9 Options 69
- 5-10 Ethically Right for Me? II 70
- 5-11 Trusting the Experts 71
- 5-12 Deception II 71
- 5-13 Confidentiality 72
- 5-14 Moral Development of Engineers 73
- Discussion Questions 74

6 Professional development 77

- 6-1 Tenure in Engineering Schools 78
- 6-2 Famous Engineers in History 80
- 6-3 Networking 81
- Discussion Questions 82

7 Solicit or accept gratuities 83
- 7-1 Deception III 83
- 7-2 Corporate Gift Policies 85
- Discussion Questions 87

8 Self-laudatory language 89
- 8-1 Advertising I 89
- Advertising II 90
- Discussion Questions 91

9 Contributions in order to secure work 92
- 9-1 Competitive Bidding 92
- 9-2 Bribery and the Law 95
- 9-3 When in Rome 97
- 9-4 Ethical Dilemmas II 98
- 9-5 Human Rights I 100
- Discussion Questions 102

10 Professional development of others 104
- 10-1 The Existential Pleasures of Engineering I 104
- 10-2 The Existential Pleasures of Engineering II 105
- 10-3 Engineering and Armaments 107
- 10-4 Vegetarianism 110
- 10-5 Reverence for Life 111
- 10-6 The Ethics of Asking and the Ethics of Giving 112
- 10-7 Maintaining the Quality of Engineering Education 115
- 10-8 Affirmative Action 116
- Discussion Questions 117

11 Overseas work 119
- 11-1 Environmental Racism 121
- 11-2 Human Rights II 123
- 11-3 Politicians and Their Reputations 125
- Discussion Questions 125

12 Uphold the honor and dignity 127
- 12-1 Manners 127
- 12-2 Workplace Harassment 129
- Discussion Questions 130

13 Faithful agents 131
    13-1 Loyalty to an Employer   133
           Discussion Questions   135

14 Avoid conflicts of interest 136
    14-1 Safety   137
    14-2 Professional Registration   139
    14-3 Conflict of Interest I   140
    14-4 Conflict of Interest II   142
    14-5 Why Be A Good Engineer?   146
           Discussion Questions   147

15 Objective and truthful manner 149
    15-1 Professional Respect   150
    15-2 Engineers and the Media   151
           Discussion Questions   152

Epilogue 154

Preface

This book began life in Aarne's yard in Chapel Hill, North Carolina, in September 1996. Aarne is your engineer author. Alastair, your ethicist author from New Zealand, was sitting in Aarne's yard, by the pool. The temperature was an unseasonable 85 degrees and he fancied a swim, but the pool was covered and locked down, and, not being an engineer, he didn't know how to remove the cover. So he decided to begin a novel about an engineer.

When Aarne got home, Alastair had written what became the Prologue. It was going to be about a Vietnam veteran who was obsessed with the designer of the DC-3, but somehow it became this book.

Along the way, we often wondered why there are so few books and courses on engineering ethics. Are there few courses because there are few suitable texts, or are there few texts because publishers, noting that there are few courses, don't want to publish texts? Do engineering faculty think students won't be interested in ethics? Is the engineering curriculum so crowded that there's no room for ethics? Or are engineers simply not interested in ethics?

All of these are genuine questions, except for the last one, to which we know the answer. Engineers are certainly interested in ethics! We have both presented ethics papers at professional conferences and meetings, to large and enthusiastic audiences. We've also found that raising engineering ethics issues in the lunchroom and the classroom guarantees a lively discussion. We're passionate about engineering ethics, and we want you to be passionate about it too—hence, this book.

By the way, Alastair never got his swim. Just as well, since Aarne had already winterized the pool.

Acknowledgments

Aaron Jarden, a former Waikato graduate student, read earlier versions of the manuscript and made many suggestions for improvement, many of which we have incorporated. Laura, Alastair's daughter, came up with the brilliant idea of writing the story in the second-person singular. Carole, his wife, read numerous drafts of the story and made lots of suggestions including the idea for the "Where are they now?" section at the end. The Sherman character is based on a New Zealand Planning Institute case that Professor Jenny Dixon of the University of Auckland gave me. Kelly Tudhope, a Waikato law and philosophy student, contributed many ideas. Ron Ziemian of the Department of Civil and Environmental Engineering at Bucknell University provided much-needed help with the structural engineering material. Ray Russell, University of Texas at Austin; Mary C. Verstraete, The University of Akron; and Dale Wittmer, Southern Illinois University reviewed the manuscript.

Alastair S. Gunn
Aarne Vesilind

About This Book: A Note to the Instructor

As you know if you've read the Preface, this book started out as a novel and ended up as a textbook, though we hope you'll find the story interesting in its own right. We have aimed it at ethics courses or modules in an engineering program, but the book could be equally suitable as an introduction to issues in engineering in a humanities program. If you plan to use it as a text (and ethicist Alastair has used an early version successfully), here are some ideas.

Most texts on professional and business ethics are anthologies of published or (less frequently) specially commissioned articles. Such books invariably begin with a chapter explaining the main ethical theories current in Western philosophy, followed by chapters on issues relevant to the profession such as relations with clients, social responsibilities, and whistle-blowing. The articles that make up most of the text are usually written by academic professionals such as philosophers for an audience of other academics.

We do not think that this is a good way to get engineering students to think about ethical issues in their professional lives. Engineering students are not more likely to become interested in ethics by reading about utilitarianism than arts students are likely to become interested in engineering problems by reading about finite element analysis. So we decided on a strategy of engaging students' interest from the beginning by telling the story of an engineer who has to deal with a range of ethical issues in professional practice. Chris, our protagonist, is a structural engineer, but the issues we discuss arise in every branch of engineering. In fact, most of them are common to professional practice in general. One of the features of the book is to show engineers who have to engage with issues that are also of concern to other professionals such as lawyers and doctors.

When Alastair used the book in a graduate applied ethics course, he handed out a chapter at a time, asked the students to identify ethical issues that arise in that chapter, had a class discussion, and then handed out the information boxes on the relevant issues. In this book we've already identified the issues, but you could ask students to note down some ideas about an issue before reading the material on, say, whistle-blowing.

Probably (we hope!) students will be so interested in the story that they will want to read to the end without going to the information boxes. You might therefore want to ask them to do this in the first week or two of the course. Then, you could assign one or more chapters (they vary in length) each week and base discussion on the issues raised in the reading. Here are some more specific ideas; most of them can be used in combination—for instance, introducing a topic as if you were one of the characters in the novel. We've used the example of the car that Chris is given in Chapter 7.

xiii

- As an icebreaker, divide the students into pairs, assigning each an opposite position—for instance, should Chris accept the car or should Joe allow Chris to accept it? They spend ten minutes on it, and randomly selected pairs are invited to report back to the group. Another way to do this is to have students in a pair report *each other's* position.
- Draw up a schedule of topics and have each student verbally introduce one (or more, depending on class size) topic, briefly presenting the issues and their own view. You could use the discussion questions as a basis for these preesentations.
- All of the ethical issues arise in situations where one of the characters has to make a decision. Should Chris accept the car or not? So, you could have the students role-play, self-chosen or randomly assigned.
- Where there are clearly only two alternative positions, divide the class into three groups. Group I believes that Chris should accept the car (or Joe should let Chris keep the car) and Group 2 believes the opposite. Group 3 is an ethics board. Groups 1 and 2 have 15 minutes to prepare their case, while Group 3 identifies relevant issues. Groups 1 and 2 present their cases, uninterrupted, and are then allowed to challenge each other's cases. Group 3 then asks each group questions, has a brief caucus, and announces its "verdict."

Finally, we would love to hear from instructors who have used this book. Please send comments to alastair@waikato.ac.nz and vesilind@bucknell.edu.

HOLD PARAMOUNT
The Engineer's Responsibility to Society

1

Doing the right thing

It's Wednesday morning, October 5, and you have driven to Sunset Beach, a quiet town on an island off the coast of North Carolina. A pleasant three-hour drive. You have to meet a client at noon, but in the meantime, you are enjoying this quaint town. You walk along the pier, looking at the eighteenth-century lighthouse, and wonder what the effect of the multistory apartment complex planned for the town will have on life in this community. But that is not your concern. Your client, the developer of the planned apartment building, wants to talk about some scheduling problems, and since your engineering firm is managing the project, you are here to accommodate him.

But you are early and work can wait for a few hours. You resist the urge to check your e-mail on your laptop. It is a beautiful, breezy, sunny day. No professional or domestic chores await your attention. You have no urgent bills to pay, contracts to write, political meetings to attend, or dogs to walk. You've had a really busy couple of weeks, and you're entitled to a few hours to yourself. You look at your watch, remembering the important meeting back in the office at five o'clock, and relax. All is under control.

You are looking forward to the weekend. You could sit in your yard and read, or go for a walk, or arrange to meet friends for lunch. You're a baseball fan, and no fewer than four playoff games are being televised. You have some decisions to make, but it's a most agreeable set of choices!

You've suggested to your client that you should meet at BJ's Seaside Café. BJ's is your favorite restaurant and BJ does great crabcakes. Then you can drive home in time to make the meeting at the office, take your daughter Laura to parents night at school, and even catch a ball game on TV.

Everything goes according to plan. The problems with the client are easily resolved, the crabcakes are delicious, traffic is light, and you are back in plenty of time for the five o'clock meeting. Later that evening, you take your 10-year-old daughter to the PTA parents night and act the part of an interested and supporting parent, to the delight and satisfaction of your daughter. You come home to watch the late game in the playoff series. Nice day!

You made a lot of decisions today; none of them involving ethical issues. You have every right to be pleased with yourself, and no one is going to begrudge you that. But let's change the scenario a little . . .

The problems with the client are more complex than you thought. The lunch at BJ's is not all that pleasant, and it promises to be protracted. The client wants to hire a specific contractor who you know to be very slow and unreliable, and this could cause serious delays in the construction. But slowly it dawns on you that the contractor is a relative of some sort, and your client is under some pressure to find the contractor work. The discussion gets a little strident, but he is your client, and you explain to the best of your ability what your concerns are. The decision, however, is his.

In an attempt to bring the conversation to a friendlier plane, you mention baseball and thankfully discover that the client is a big Atlanta Braves fan, as are you, and he is impressed by your knowledge of baseball. You get deeper and deeper into discussions on the finer points of the national pastime—you describe how you actually saw the infield fly rule applied at a major league game—and by now you are on your fourth (fifth?) glass of chardonnay. Before you know it, it's four o'clock.

You excuse yourself for a moment and call your office on the cell phone. The secretary answers.

"Rosemarie?" you hear yourself say, trying hard not to slur your words. "I got big problems here. It's important that I spend some time with our client. Please tell Joe and Sarah that I can't come to the marketing meeting. Yes, I know it's really, really important, but this is critical too!"

You did not represent the situation accurately, of course. You could easily have excused yourself at two o'clock and made it back in time, but you were having too good a time and the conversation was too interesting.

Box 1-1

Deception I

Moral values such as truth, honesty, duty, caring, and others define the society in which we would like to live. If everyone within this society, or *moral community*, abides by these values, then we are all better off. We all benefit for having all of us be moral people.

One of the most important moral values is truthfulness, or in the negative, lying. People can lie in many ways: by stating to others something they know to be false, by body language and other signals indicating something that is not true, or by *deception*. The latter is the withholding of information that leads the listener to form incorrect conclusions.

Suppose, for example, an engineer chooses to specify a water pump for a water treatment plant. The client assumes that the engineer has made an informed and disinterested decision. But suppose the company that manufactures the pump is owned by the engineer's brother-in-law. If the client is aware of this, he might question just how disinterested the engineer really is. Maybe the pump is a piece of junk and the engineer is doing his brother-in-law a favor. This may not be the case at all. The pump might be the best there is and can be bought at a reasonable price. But the deception occurs when the engineer does not tell the client, who assumes that there are no conflicts of interest in the purchase of the equipment. The client should not have to ask every time, "Is the supplier your brother-in-law?" It is simply assumed that the supplier is not, and thus to withhold this information is deception.

In the scenario, Chris deceived Rosemarie by not stating that the afternoon was spent discussing baseball. Rosemarie simply assumed that the discussion was on the project. If Rosemarie had asked Chris, "Have you been talking baseball for the past two hours?" then Chris would have been in a position of having to either lie or tell the truth. By not making clear what had actually happened, Chris clearly deceived Rosemarie (and indirectly Joe and Sarah, the managers in the engineering firm who expected Chris to be at the five o'clock meeting).

You also call Alex's work number, but nobody answers. You leave a message saying that you will be late and to not wait dinner. Wonderful, understanding Alex, who puts up with your crazy professional schedule, takes care of the home front, and often has primary responsibility for your two children. What would you ever do without such a saint-of-a-person?

Your client suggests that you repair to the bar and continue the conversation. The Atlanta-Colorado playoff series game is on, and you get sucked into watching Glavine outpitch Lee Kim. Suddenly, it's seven thirty. The game is over.

With a jolt, you remember that you had promised Laura to take her to the PTA parents night at school. You call her, and she answers the phone. You start to explain why you can't come home and she breaks out crying.

"But you *promised* me!" she sobs.

"I know, sweetheart, but I just couldn't possibly get off from work in time. Let's do something together this weekend and make up for this," you say, hoping that she will be mollified. You feel rotten for having disappointed Laura.

Box 1-2

Keeping Promises

One of the basic rules in Western societies is that we should keep promises. Some philosophers call such a moral rule a *prima facie* moral rule, meaning that it should generally be followed, unless there are strong reasons for breaking it.

Consider this example: You are driving home, having promised your parents to come to dinner at six o'clock. You are about to leave when a friend of yours becomes very ill, and you are the only one who can take her to the hospital. You drive her to the hospital, and now you cannot make it to your parents' house for dinner. You have broken the promise to your parents, but everyone (including your parents) will probably agree that in the circumstances your good deed clearly overrode the promise, and you did the right thing.

But now consider another scenario. You dawdled at work, playing solitaire on the computer, and did not leave on time to make it to dinner at six. You know you are late, so you speed, going 85 mph in a 55-mph speed zone. You argue that it is important to not break a promise to your parents, and you are speeding in order to make it in time and to keep your promise. Is this action morally supportable?

Most people would probably say that it is not, because instead of doing something positive such as helping another person, you are creating a potentially dangerous situation that could cause great harm to others (you could cause a serious wreck). The promise to your parents is important, but in the circumstances that you have created it is no longer important enough to allow you to speed and possibly hurt others.

In Chris's case, the promise to be home for the PTA meeting was broken not by some deed that resulted in greater good, or even by some other commitment (such as the need to do work), but rather by simple self-indulgence. You probably agree that Chris has every right to feel rotten.

Your client decides to go on home, and you are stuck with the bill for dinner and drinks. You ask the bartender to add things up for you and get ready to leave. Instinctively, you look at the total and do a rough calculation of what it ought to be. The bill is much less than you expected. A closer scrutiny reveals that the bartender has made a mistake in adding up the dinner and drinks. Instead of the bill for about $120 that you expected, it is only $58. You point this out to the bartender, who realizes that he has forgotten to add all your drinks to the bill. He is very grateful to you for pointing this out, since the mistake would have come out of his already meager salary.

Box 1-3

Doing the Right Thing

Moral values reflect how we *ought* to treat each other. One can do many dumb things, like getting drunk, but if this does not harm others, it is not of moral concern. Ethicists often distinguish between actions that harm others and actions that harm only the agent. Concerns for the latter are often referred to as *prudential* rather than genuinely *moral* considerations. The well-known British utilitarian philosopher, economist, writer, and politician John Stuart Mill (1806–1873) illustrates it this way, in his book *On Liberty:* An engine driver (locomotive engineer) ought not to be drunk at work, because he would be endangering the lives of others, but if he chooses to drink himself into oblivion every night at home, he harms only himself. His conduct in his private life may be stupid and unwise, but it is not *wrong*.

Morality is not only about not doing wrong things; it is also about doing right things. Chris had a chance to walk away from the erroneous dinner and bar bill. If the mistake had been caught before leaving, Chris could easily have feigned ignorance. But moral behavior asks us to think about how we ourselves would like to be treated, and Chris's honesty in paying the full bill was very much appreciated by the bartender.

In our culture, the impetus for doing the right thing comes from two sources: religious and secular.

In many religious traditions, notably Christianity, doing the right thing is doing what is commanded by God. Christians believe that what is commanded by God can be found in the Bible, which many believe to be the literal word of God. Following the Ten Commandments is thus a manifestation of doing God's will, and actions that follow these commandments are thus defined as moral actions.

One of the problems with this approach to morality is that first one has to believe in the existence of God, and then that God's will is known. Through the ages, people have believed in various gods, often defining god as the gap between what we can see and what we know. For example, the Northern Europeans had a god of lightning. They could see the lighting bolts and witness their destructive power, but they had no knowledge of electricity. This gap was filled by a god, and this explained the phenomenon of lighting. In our world, death is a mystery, and yet we see people die. The mysterious gap is filled in many monotheistic religions with the belief in a heaven, with hell for good measure for not following the rules of the religion. If gods exist in gaps, it is unclear just which god we are invoking when we base morality on God's commandments. Each religion would have its own set of rules for what is right and what is wrong.

Another problem with using the word of God as the source of morality is that we have difficulty knowing what it is that God wants us to do. If the Bible is taken literally as the word of God, then there are a lot of commandments given in the book of Leviticus that we would find irrational or just plain silly, such as not being allowed to have round haircuts (Leviticus 19:27). The God of the Bible has very little to say about the cloning of humans, or birth control, or nuclear weapons, or engineering. In fact, the God of the Bible is a fairly destructive God when it comes to engineering, destroying the Tower of Babel, for example, which must have been an amazing engineering feat.

And finally, the problem with using God as a source of morals is that there is no logical reason for accepting one god over another god. If a person believes in a god such as the Christian God, then that is (happily for us in the free world) a very personal decision, and we would respect such a choice. We would only ask that people with certain beliefs in their personal god not ask the rest of us to believe likewise. And without such belief, whatever moral systems they have developed based on their religion would not be shared by others.

A counterargument to the above would be that this is all nit-picking. The main idea of most of the world's religions is the same—a caring for each other and a sense of community under the protection and watchful eye of God, however defined. There is much to recommend this view. Many of the best people (in the sense of acting morally) are people with deep religious beliefs, and it is obvious that these religious beliefs are central to their understanding of right and wrong. But it is unclear if the goodness of these persons came because of their religious beliefs, or if they were good people who embraced a religion that then supported their worldview and personal human interactions.

(continued)

Box 1-3 (continued)

If Chris were a Christian, the decision to tell the bartender about the undercharge might have been based on religious beliefs. Walking away without paying might have been breaking one of the Ten Commandments. More important, Chris would not have demonstrated the caring and love toward others that Christianity stands for. A good Christian would have paid the bill.

The second source of morality is totally secular. It comes from the notion of the societal agreement that we make with each other to live our lives so as to be good to each other. This idea was first articulated by Thomas Hobbes (1588–1679), who called it a "social contract." We agree to behave in certain ways toward each other, and if we do, then we all will benefit from it. For example, if we agree not to steal from each other, it will be a much better world to live in than one in which people cannot be trusted not to steal. Just imagine for a minute a society in which you could lay nothing down for fear of the next person taking it. There are such societies in this world (and even in the United States), and we would be very much the poorer for it if we allowed this to happen in our own communities. In fact, we feel so strongly about stealing property that we have made laws to punish people who steal. The idea of the social contract is to try to maintain a society in which people will respect each other and follow a set of moral rules, thereby enhancing the quality of life for all.

The problem with this idea is that there is no such thing as a societal contract. Or as one ethicist put it, "it is not worth the paper it is not written on." If we do not agree on the set of moral values we would like to abide by, then all bets are off, and we are in a society where every person is out for himself or herself, and taking all you can from others, not trusting others, not believing others, and not caring for each other is the norm.

The societal contract is almost impossible to implement when moral behavior in a society has broken down. Trying to convince each person to behave in a moral way will then get the response "Why should I behave in this way when nobody else does?" This is a difficult question to answer. What we would like, therefore, is to prevent a society from ever getting to the stage of "every person for himself or herself" because once it is there, it is very difficult to recover and bring that society back to a more moral lifestyle.

If Chris in our story believed in the social contract, and believed that it was the responsibility of every person to be honest (so that we all would benefit), then this would have explained the action of correcting the bar bill.

Thus, it is of little importance where our moral values come from. What is important is that we agree what these moral values are, and then agree to abide by them.

You pay the bill, take stock of the situation, and decide that you are in no shape to drive home. Luckily there's a motel next door—you could stay there and drive home in the morning. But the motel is a five-star resort, offering rooms starting at $250 per night. Too much, you think. What other choices do you have? You don't want to sleep in your car—too dangerous, not to mention uncomfortable.

You decide to drive homeward and stop at the next reasonably priced lodging with a vacancy sign, but there are no motels. Anyhow, you feel much more in control now that you're behind the wheel. Why not keep going and just drive carefully?

You are sleepy, and it is a long, boring drive. You slip a CD into the dash and hear Frank Sinatra singing about how much he likes New York, New York. You always get a kick out of Frank Sinatra, a man who lived a full, if not necessarily ethical, life. The CD, you recall, was given to you by the accountant who keeps the books for your engineering firm. He knew you liked Sinatra recordings, and downloaded it from MP3 and burned the CD for you. Every time you listen to it you have a vague notion of having done something underhanded, but you can't put your finger on what bothers you. It is free music, on the net, and the accountant burned the CD for you, and besides, Sinatra is dead. Is it the "free" music that bothers you, or is it how you got the CD? You originally recommended this accountant for keeping the firm's books. Was it ok to accept the CD from him? You decide not to think about it and sit back and enjoy Frank singing about how much he likes Chicago, Chicago.

Box 1-4

Theft of Music

One of the moral rules in our society is "Do not steal." In our society (though not in societies where rights are not the primary vocabulary of morality), this rule is based on the view that humans have a right to private property and that this right should be protected by society. Private property is protected in many religious traditions, for instance in the Old Testament and in the Koran.

In the case of the Internet, copying material for one's personal use is permitted and even encouraged, so it is not possible to steal things from the Internet. Some information on the net is for sale, such as term papers, and there is nothing immoral about purchasing such term papers. These papers are not illegal or stolen property, and the transaction is perfectly appropriate. No moral rules are being broken. The moral infraction occurs when and if these papers are submitted for academic credit. In this case, the student is guilty of deception (see Box 1-1). The professor does not need to ask "Did you write this yourself?" The assumption is that the work is by the student.

There is some question about the moral health of those people whose business it is to sell term papers on the Internet. They must know that what they are doing is causing students to cheat, or at least creating conditions for promoting cheating. There is little difference (in kind) between people who sell term papers and people who sell cocaine. Both can argue that they are not doing anything immoral since it is still up to the purchaser to choose to use or not use the product.

We could then argue that if we download material from the Internet solely for the purposes of our own entertainment and not for credit or for profit, then there is nothing wrong with this. Legally, one is able to photocopy copyrighted materials without permission if the copies are solely for personal use. Why not then be able to download music if it is only for entertainment and not for sale?

The argument against this is that the music is rightfully the property of the entertainer, so he or she is entitled to receive financial compensation for the work. If anyone is wrong here, it is the people who *upload* the music on the Internet, just as the wrong (if it is wrong) is in the people who upload the term papers.

"Pirates" argue that making copies available to the public is fair game and is not immoral, nor should it be illegal. They also say that only a minimal proportion of the income that artists receive comes from CD sales. The CDs should be used, they argue, as introductions to the entertainers so that people will go to their concerts.

What do you think? Is it immoral—and should it be illegal—to download music from the Internet, solely for one's own entertainment?

Halfway home, you are startled and horrified to see an 18-wheeler about half a mile in front of you suddenly jackknife and spin off the road into a ditch. You pull up behind it, get out, and find that the driver is conscious but apparently unable to move. You have a first-aid certificate and you'd like to help him, but you don't want to be around when the Highway Patrol arrives. You're almost certainly over the legal limit of alcohol, and the last thing you need is a DUI conviction. You have a cell phone in the car, but you don't want to dial 911 because they will be able to trace the call. So you drive to the rest area that, you remember, is only a couple of miles further along the interstate, make an anonymous call to the state police, and drive home without further incident.

Box 1-5

Obligations to Strangers

Is there an obligation to help those in need? Sometimes you have a legal obligation to do so: You are responsible for providing assistance due to your relationship with other persons such as your children, your patients, or the passengers in the bus you are driving. What about obligations to *strangers*, as in this case?

Most people are familiar with the biblical story of the Good Samaritan, who helped a Jewish stranger who had been attacked by thieves and left for dead. The Samaritan, who happened to be passing, provided help even though Israel and Samaria were traditionally enemies. Jesus told this story as an explanation of the obligation to love one's neighbor, in response to a person who had asked, "Who is my neighbor?" The point of the story is that one has a moral obligation to be altruistic—to come to the assistance of anyone who is in need, even at some cost to oneself and even if there is no socially accepted obligation to do so. This principle is found in many religions including Buddhism and Islam.

The term *Good Samaritan*, however, has come to mean a person who comes to the aid of another when there is considered to be no *duty* to do so. We may admire and respect people who go out of their way to assist total strangers, but we would not necessarily feel under any obligation to do so ourselves. Some ethicists use the term *supererogatory* to describe such behavior—a Latin-based term meaning "going beyond the call of duty." It is praiseworthy to go beyond duty, but no one can be blamed for not doing so. Thus, in some jurisdictions there is a legal obligation to be a Good Samaritan. The state of Vermont, for example, passed legislation to this effect in the wake of the murder of Kitty Genovese in 1964, when some 38 people witnessed or heard the killing but no one made any attempt to come to her aid or even call the police.

Peter Singer, in his well-known book *Practical Ethics* (1993), argues that we have an extensive obligation to help others who are in need, including an obligation to donate to famine relief rather than spend our resources on nonessential items for ourselves and our families. He bases this claim on the principle that if we can stop something bad from happening, without excessive cost to ourselves, then we ought to do so. The utility gain to people who are saved from starving is clearly greater than the utility loss to the donor, who merely loses some material comfort. Therefore, we have an obligation to donate to famine relief to the point that, if we donated more, we would ourselves be in want. To help his readers to understand this principle, he uses the analogy of a person who notices a small child in danger of drowning in a lake. Even though the child is not yours, he argues, you surely have an obligation to save him or her, even at some cost and inconvenience to yourself.

Opponents of Singer's position, for instance Garrett Hardin (1978), argue that such donations will bring about a larger population and thus lead to even larger-scale famine. Keeping people alive so they can procreate and further increase the population in countries in which the agriculture cannot support the existing population is simply immoral. Others argue that no one has an obligation to donate any of their lawfully acquired property to anyone (Hospers 1995), or that, at most, a person may be said to have an obligation to donate their "fair share," that is, the amount that would alleviate famine if everyone made an appropriate donation.

In this version of the story, you have made a number of decisions that obviously have ethical dimensions. Most people would probably disapprove of the lie to the secretary and breaking the promise to attend the PTA meeting, approve of helping out the bartender by catching his error, and agree that you behaved wrongly both in driving after drinking and in not helping the truck driver. Owning a CD that was downloaded from the Internet might be ambiguous, and there may be differences of opinion on whether or not you behaved dishonorably by accepting the gift from the accountant.

> ### Box 1-6
>
> ## Moral Rules
>
> One way of thinking about what would be the right thing to do for a given situation is to have a set of rules handy that can be used for evaluation. Dartmouth College moral philosopher Bernard Gert (1988) believes that rational, impartial persons would agree on a common code that is to everyone's advantage, and that it is rational to be moral. He believes that it would be irrational to reject this code or to break the rules without adequate justification. Gert's proposed new "ten commandments" are these:
>
> 1. Don't kill.
> 2. Don't cause pain.
> 3. Don't disable.
> 4. Don't deprive freedom.
> 5. Don't deprive pleasure.
> 6. Don't deceive.
> 7. Don't break promises.
> 8. Don't cheat.
> 9. Don't disobey laws.
> 10. Don't fail to do your duty.
>
> As Gert recognizes, to make such a list invites the objection that one may be in a situation where several of the commandments conflict—as we note in Box 1-2. Suppose you are late for an appointment. Should you speed in your car (disobey the law) or be late (break a promise)? Gert suggests that we ought to follow these rules unless rational, impartial persons would agree that one of them should be broken. For example, if most rational, impartial persons would agree that wanting to watch the baseball game was sufficient reason for breaking a promise to Chris's daughter, then this would be the right thing to do.
>
> Gert argues that a rational, impartial person who agreed with this list of moral rules would say that Chris should have kept the promise. But the problem is that "rational, impartial persons" do not come with universally accepted identifying features or labels. Who are these rational impartial people to whom we should defer decisions on moral behavior?

Decisions requiring moral reasoning occur not only in everyday life but also in professional life. Many such moral choices, both personal and professional, are made by the main character in the story that follows. How Chris resolves such dilemmas is the essence of this story, and of professional ethics.

Discussion Questions

1-1. Concerning famine relief, who has the stronger argument, Singer or Hardin? Why?

1-2. Suppose you are careless with your driving and wreck your car. You argue that society should pay for your new car because the cost would be distributed evenly over the entire population so that no single person would ever feel the cost, but the benefit to you would be substantial—you get a new car. How is this scenario different from feeding a starving person, or would the results be the same?

1-3. Suppose you promise to meet your friend for lunch, and you just plain forget. She confronts you the next day. You decide not to tell her that you just forgot, but to make up an excuse. List some excuses you might make and analyze each in terms of whether or not the excuses are (a) believable and (b) truthful. Which excuse would you have used if this had happened to you? If none, explain why you would choose to tell the truth.

1-4. Some people would excuse Chris's lie to the secretary as a "little white lie" because it harms no one. Do you agree with this judgment? Are such lies acceptable?

1-5. Based on what you know at this point, what do you think of Chris's character? Is Chris an honorable person? Why do you think so or not think so?

1-6. What do you think of the argument that downloading music is acceptable because artists make very little money on it anyway? Is this a legitimate moral argument? Why or why not?

References

Hardin, G. 1978. *Limits to Altruism: An Ecologist's View of Survival*. Bloomington: Indiana University Press.

Hospers, J. 1996. *Human Conduct: Problems in Ethics*. Belmont, CA: Wadsworth.

Singer, P. 1993. *Practical Ethics*. New York: Cambridge University Press.

Some good short books on the question of morality and ethics, written for the non-philosopher:

Gert, B. 1988. *Morality*. New York: Oxford University Press.

Rachaels, J. 1989. *The Right Thing to Do*. New York: Random House.

Rachaels, J. 1995. *The Elements of Moral Philosophy*. New York: Random House.

Weston, A. 2001. *A 21st Century Ethical Toolbox*. New York: Oxford University Press.

2

The engineering profession

Sunday, October 6

You sleep in until nine, shower, feed Mango, the cat, and drive to Mama Kit's Kitchen for breakfast. While you're waiting for your pancakes, who should walk in but your old friend Kelly. You were at grad school together, but you haven't seen her for ages. The years in California have taken a certain toll on her fashion sense, you think, but she looks great with her deep suntan, long blonde (apparently still natural) hair, ethnic dress, and Mexican silver jewelry. You're delighted to see each other, and you look forward to catching up with all her news. It turns out that, after graduating, she moved to California, worked for a few years, got married, got divorced, did a law degree, worked for a firm in San Diego, and recently got a job with Allegheny Environmental Services, just a few miles away from your company.

This all sounds fine to you, but Kelly doesn't look particularly happy. After a while, she tells you that she's having a major problem at work. Her job mostly involves representing Allegheny's interests with state and federal environmental agencies.

"We reprocess these residuals from this factory in New Jersey, the plant releases a lot of chromium. I won't bore you with the details, but like there's two main kinds of it, Cr III and Cr VI. Cr III isn't toxic, no problem, but Cr VI is a *lot* toxic, and that's what our waste mostly contains. But our NPDES[1] permit only specifies total chromium, and we're within that."

"So you're not violating state law?"

"No. I'm supposed to be in charge of this subsidiary's environmental program, what a joke. I told Arthur, my dweeb boss, it's polluting the river, never mind the law, and he agreed, but then he goes, 'Don't worry, it's not a technical violation. We picked up this company for a song, and we plan to run it with the existing obsolescent plant for three years and then upgrade.' He talks like that, all the time."

"Still got your eidetic memory, Kelly?"

"Totally, course I do."

"Why do they want to keep using the old plant?" you ask.

"Hey, it looks like an East German soda ash plant. But Arthur says it's making money. He's like, 'We're going to replace it in a few years. Listen, Kelly, the way this state works, if the EPA decides to require us to test for Cr III and Cr VI separately, which my guess is it will, the worst that can happen is we'll have to bring the upgrade forward by a year or so,

---

[1] NPDES is the National Pollution Discharge Elimination System, and the permits are granted to all those industries and municipalities that discharge treated wastewater into the waterways.

and even then we'll probably be able to stall them, so it won't cost us anything we weren't going to spend anyway.' "

"Well, what do you want to do?"

"If you put me in charge, I'd close it down if I had to. But it's not, you know, what I *want* to do, Chris; it's what I *ought* to do. I'm a professional, I totally need to do the best job I can, and that doesn't mean dumping Cr VI in the river. But they pay me to run it their way."

Box 2-1

On Being a Professional

When Kelly says, "I am a professional," what is she referring to? What difference does it make if one is or is not a professional?

The word itself has an interesting history. During the Middle Ages, most of Western knowledge was kept alive by the monks who lived in the monasteries. These men had "professed" their faith and became "those who profess," or "professers." Today's professors are in a direct lineage to that tradition. Eventually, everyone who possessed specialized knowledge that required education and study became known as a "professional."

There is, of course, a second common meaning to the word *professional*: one who gets paid for a service. Thus, we have, for instance, amateurs and professionals in sports. But this is not what Kelly means. She is referring to a special obligation that professionals have to serve the needs of the public.

Although there are many lists of criteria for what is and is not a profession, the following is widely accepted. A professional is one who

- Has extensive tertiary training and/or education
- Practices an art that requires significant intellectual development
- Provides an important service to the public
- Is certified or licensed by the state
- Has an organization that practices self-regulation and controls entrance to the field
- Receives power from the state in return for a commitment to the public good
- Belongs to an organization that has a code of ethics

Using this list as a guide, certain vocations are clearly not professions. Barbers, for example, provide a useful service to society and in some jurisdictions are licensed by the state, but they do not have extensive training in a field that requires significant intellectual development. The fields that clearly meet these criteria are medicine, law, theology, engineering, nursing, accounting, pharmacy, and perhaps a few others. Among those that do not, somewhat surprisingly, is college teaching—most professors have no training in how to teach.

Perhaps the most interesting feature of professions is that the state gives special power to professionals. In medicine, for example, physicians are empowered to perform certain actions (such as surgery) that are not permitted by others. Pharmacists, likewise, are permitted by the state to dispense drugs—an action that would land the rest of us in jail if caught. Similarly, engineers are licensed, exclusively, by the state to seal engineering plans and specifications. More on this later.

You, for the umpteenth time, reflect how grateful you are that your job does not require you to be beholden to your employer. As a consulting engineer, you retain your independence and work for the benefit of the client. You also have the autonomy, theoretically at least, to take whatever job you want and reject those you do not want. Society in fact *expects* engineers to be critical and choose not to do work that would be to the detriment of the public.

Box 2-2

Technical Expertise and Ethical Obligations

Engineers are technology experts, unlike most people who have no idea how the technology that they use every day actually works. For instance, they drive their cars, turn on their televisions, work in high-rise buildings, use their computers, and draw out money from automatic teller machines, but they have no idea how these systems work and rely totally on engineers to ensure that they do work. Of course, no engineer can be an expert on everything, so engineers themselves rely on other engineers. The difference is that for an engineer, technology is never mysterious. Any engineer can quickly explain a highly technical problem in his or her own specialty to any other engineer. The same can be said for lawyers, doctors, and other professionals.

Kenneth Alpern (1983) argues that this situation creates a special obligation for engineers, even to the extent that they may be called upon to be moral heroes, and engineers have a duty to make personal sacrifices in calling attention to defective design, questionable tests, dangerous products, and so on. "Engineers are bound by no special moral obligations, but ordinary moral requirements stipulate that the engineer nonetheless be willing to make greater personal sacrifices than can normally be demanded of people in general. This qualifies engineers as moral heroes of a certain sort . . . specialization of labor and the delegation of authority characteristic of hierarchical organizations does not insulate the engineer from moral responsibility nor from the requirement that he be prepared to exercise moral courage."

In other words, in certain circumstances, engineers have a positive duty to whistle-blow (see Boxes 5-6 and 5-7). When Kelly says that she is a professional, she is expressing the central ethical problem in engineering. If she exercises moral judgment, she will probably lose her job. She believes that what she is doing is not right, but she feels constrained by her job responsibilities, as well as her obligations to her family.

But you also know that many engineers, like Kelly, do not have the privilege that engineers have historically had—of being autonomous professionals. Too bad that U.S engineers had not been smart enough to concentrate their influence into one professional organization, but instead ended up with separate professional organizations for each subdiscipline.

Box 2-3

Organization of Professional Engineering

Professional engineering has its roots in building whatever people can use. The earliest engineers built walls and buildings and means for defending cities. Other engineers developed offensive weaponry such as catapults (the engines of war). Throughout history, engineers were special people because they could get things done.

From these builders there emerged a special set of professionals who prided themselves not so much on getting things done as in visualizing how things should be. Today we call these professionals architects, and they differ from engineers in that they are first and foremost artists. They don't get involved in the technical minutiae of why buildings stand up or how they are built. Form is as important as function to architects, and they often point to truly ugly buildings designed by engineers who considered only function.

(continued)

Box 2-3 (continued)

The engineer builders may have evolved to a social level not much different from that of the other manual arts such as masonry and carpentry were it not for the architects' abhorrence for anything numerical. What made the engineer truly useful to society was the ability to calculate whether a building or a bridge will or will not fall down, or whether a boiler will actually hold the pressure it needed, or whether water would actually flow at the required velocity. What elevated engineers to the status of professional was the incorporation of higher engineering skills, particularly a knowledge of calculus. With these tools, engineers became highly educated and were able to accomplish previously unimaginable feats.

In the United States, the first engineers were often self-taught men with little formal education. George Washington, for example, taught himself how to survey. The first engineering education in the United States was at the Military Academy at West Point. During the Civil War, the engineering education at West Point was an obvious advantage on the battlefield, for both sides. Following the example of West Point, the first nonmilitary engineering school was established at Troy, New York, as the Rensselaer Polytechnic Institute. Soon other private engineering schools were founded, including Cooper Union, Stevens, and Lehigh. Then in the 1860s, the land grant colleges opened, providing a steady supply of educated engineers. Because these engineers were engaged in public (civil) work as opposed to military work, all nonmilitary engineers became known as *civil engineers*.

The earliest engineering societies were mainly social clubs, but by the latter part of the 19th century, these organizations began to have an effect on engineering practice. By the early 1900s, engineers had many feats to their credit (for example, the Brooklyn Bridge) for which they were highly respected. The two most influential engineering organizations were in Boston and New York, with the latter eventually becoming the American Society of Civil Engineers (ASCE). Concurrently, a group of entrepreneurial craftsmen were developing the basis for mechanical engineering and wanted to become members of the engineering society. Stupidly, the civil engineers did not welcome these skills into engineering—skills such as building steam engines and other machines. Thus was born the American Society of Mechanical Engineers (ASME). These exclusionary practices by the civil engineers eventually resulted in the founding of the American Institute of Chemical Engineers (AIChE), American Institute of Mining Engineers (AIME), several electrical and radio engineering societies that finally merged to form the Institute of Electrical and Electronic Engineers (IEEE), and a number of others.

Periodic futile efforts are made to unite the engineering societies to form a single organization that can represent the entire profession, much like the American Medical Association (AMA) represents the medical profession, the American Bar Association (ABA) represents the lawyers, and the American Institute of Architects (AIA) speaks for the architectural profession. None of these efforts have succeeded, however.

Another approach at uniting the engineers came in the development of a brand-new organization that was designed to represent the entire profession—the National Society of Professional Engineers (NSPE). Few engineers bother to join NSPE, however, and it has little influence in speaking for the profession.

Partly as a response to the conservatism of the older engineering organizations such as ASCE, IEEE, and ASME, NSPE created the National Institute of Engineering Ethics (NIEE). Although this organization was originally within the National Society of Professional Engineers, it became independent in 1996. NIEE publicizes codes of ethics and conducts ethics workshops in the United States and internationally. For example, in 1997 and 1998, with National Science Foundation assistance, a team conducted workshops to assist Russian educators to develop methods of teaching ethics to Russian engineering students (from NIEE's online newsletter, *Engineering Ethics Updates*, www.niee.org/may_98.htm).

You tune back to Kelly and hear her say, "And another thing, Arthur reminds me I'm halfway through my six-month trial period. He goes, 'We're really pleased with your work so far. I hear you ran rings around the State at the Coastal Commission hearings on the Snow Goose Resort project, and you managed to do it without making them look like the damned fools they are. Why, the assistant coastal commissioner called me just the other day to say how impressed they were with you. Said if only he had the budget, he'd hire you away tomorrow.'"

"So your future with Allegheny might depend on—"

"You got it. And with Mom in that nursing home—it's nice but if I lose this job I just really can't pay for it. And she and Dad helped me through ten years of school, I truly owe them. I can't afford to think about ethics at this point."

Box 2-4

Can We Afford to Be Ethical?

The anonymous author of a well-known article, "I Gave up Ethics—To Eat!" (Anon 1957), tells the story of how he graduated from engineering school with a commitment to high ethical standards. He worked for a number of years for an established engineering firm and left to set up his own practice. "Without question I was qualified technically. . . . I thought that with my background and willingness to work, I would soon, before long, be making a good living from my practice. Instead I soon was starving." After two years with no work, a friend tipped him off: "Nobody, he explained, gets engineering work without practicing the fine art of 'political engineering.'" He discovered that practices such as the use of a commission agent and competitive bidding, practices that are unethical according to the Code of Ethics, were required in order to obtain engineering work in his field of public works. He was aware that the agents he employed, and to whom he paid 10% commission on work that they obtained for him, were probably engaged in illegal practices such as payoffs to public officials, but "I had no way of knowing what my commission agent did. I made it a policy to give him a straight commission, and how he managed his end of the business was entirely up to him."

The anonymous author effectively operated under "two codes of ethics": As an engineer, he always worked to the highest professional standards. However, as a business person, he had no option but to act unethically as, he claims, most others do.

"Have you looked at the ASCE Code of Ethics recently, Kelly?"

"Code of Ethics? No. But from what I remember, it's so vague . . ."

Box 2-5

Engineering Codes of Ethics

Kelly ought not to have dismissed the codes so quickly. In some circumstances, the codes of ethics can be very helpful when engineers have to make value-laden decisions. This is one of the major reasons for having such codes.

Every profession must determine how it is to interact with fellow practitioners, with clients and employers, and with the public at large. Almost always, these rules are set out as a code of ethics.

"Code of ethics" is a misnomer, however. Ethics is the process of careful deliberation of the right and wrong thing to do in a given circumstance, and ethics cannot therefore be reduced to a code. What are commonly called codes of ethics are actually lists of guidelines that spell out to the practitioners and the public what the responsibilities of the individual professionals are. As such, a better title for codes of ethics might be *guides to responsible conduct*.

Engineering as a profession saw a stepwise development in its codes of ethics. The earliest codes addressed ethical concerns between and among fellow engineers and included such rules of conduct as "do not steal another engineer's client" and "do not speak

(*continued*)

Box 2-5 (continued)

disparagingly about a fellow engineer." With time, the codes incorporated the engineers' duties to employers and to clients, and such rules as "be loyal to your employer" were added. Then, in the 1970s, the recognition that engineers held a moral responsibility to the public led to the addition of the famous "engineers shall hold paramount the health, safety, and welfare of the public" clause.

In engineering, the typical code of ethics has two types of statements that reflect morality: admonitions and requirements. Admonitions are statements that strive to lead the engineer to the moral high ground, to make the engineer design his or her professional life so as to routinely act with moral integrity. Admonitions are statements specifying what the engineer ought to do to be a good engineer. To not adhere to an admonitions statement in a code of ethics will not get engineers in trouble, but to adhere to the admonition will make them better engineers. For example, in the ASCE Code of Ethics, Guideline 7a reads:*

> 7a. Engineers should keep current in their specialty field by engaging in professional practice, participating in continuing education courses, reading in the technical literature and attending professional meetings and seminars.

Nothing bad will (probably) happen if the engineer ceases to learn, but a better engineer is one who keeps up-to-date. The reason this is a *moral* statement is that engineering involves the public, and not keeping up with technical developments could result in an incompetent design that may harm the public.

A code of ethics is also full of *requirements*, things the engineer *must* do to continue to be part of the engineering community. The difference between admonitions and requirements is that not following the requirements *can* result in harm to both the engineer and the public. Ignoring the requirements would be to act immorally in professional engineering.
For example, Guideline 1b states:

> 1b. Engineers shall approve or seal only those design documents, reviewed or prepared by them, which are determined to be safe for public health and welfare in conformity with accepted engineering standards.

If a structural engineer decides to curry favor with an industrial client by approving a truss design that does not quite meet the steel design standards, the engineer is acting immorally. If the truss should fail, people might be hurt or killed. The engineer then is at fault and would suffer serious consequences.

"But one of the reasons for the codes is to help engineers make good decisions."

"That may be, but I'm a lawyer now, not an engineer. I doubt if the ASCE Code of Ethics even applies to me any more."

Box 2-6

Can a Person Stop Being an Engineer?

Probably all engineers are familiar with Scott Adams's popular *Dilbert* comic strip, which appears in over 1400 newspapers around the world. (If it isn't in your newspaper, you can access it on the Internet at http://www.dilbert.com.) *The Dilbert Principle* (the title of Adams's 1996 book) is that "the most ineffective workers are systematically promoted to the place where they can do least damage: management" (p. 14).

Unfortunately, as Adams explains, "This has not proved to be the winning strategy that you might think."

Dilbert, unlike his creator, is an engineer whose attempts to practice in a professional manner are constantly frustrated by a combination of the latest fad management theory and his arrogant, stupid, pointy-haired boss.

Managers themselves sometimes believe that engineers have no understanding of organizational and commercial realities and are motivated by a constant desire to goldplate all girders and build in at least four redundant systems.

(continued)

---

*Sourcce: Excerpts here and throughout are from the ASCE *Code of Ethics*.

> **Box 2-6** (continued)
>
>
>
> Can a person ever stop being an engineer? Certainly you can stop practicing, but are you still covered by the Code of Ethics, regardless of your professional status? For example, suppose that an engineer decides to chuck it all in and goes to Maine to retire. She finds that a local engineer had designed a wastewater system that will harm the best fishing stream in the county. It just won't work as designed, and the engineer knows this. The Code of Ethics states that an engineer, as a knowledgeable person, has a responsibility to prevent a bad situation from occurring. But she is no longer registered as a Professional Engineer. Does she still have that responsibility to do something?
>
> To take another example: Suppose that a retired physician who has ceased to actively practice is on an airplane and someone has a heart attack. Is he responsible for assisting the patient? Or would rendering assistance be praiseworthy but supererogatory? (See Box 1-5.)
>
> An engineer who "graduates" to manager may no longer be practicing as an engineer and may not maintain her professional engineering license. Our own view is that she is still an engineer. It is like a tattoo. It won't wash off.

You are on unsure ground here, so you decide to make light conversation.

"Tell me more about Allegheny. What happens if they still like you after the six months is up?"

"Then it's cool. Allegheny hires everyone—like, janitors to president—on trial. Then, if they want you, you got a job for life. They say we never lay anyone off, we only hire the best people, and once we decide you *are* the best people, you get to work for us as long as you do your stuff. Honest, apart from their environmental policy they're a cool company to work for—they even got listed on that new Internet site, you know, America's Best Employers for Professionals."

"Yeah, You Figure is also listed," you say, referring to the nationwide company that specializes in cosmetic body surgery. "And they're fighting a two-billion-dollar class action suit after those three people died."

"For sure, but if I stay with Allegheny—and I hope to—I can work on their environmental policies. Office gossip is, Arthur's going out on his own in a year or so, and the new guy might have better professional standards. And I'm like . . ."

"It might be you."

"Bitchin'. And even if it isn't, then maybe they're not the right place for me, and I'll be on my way to move to a better job with another company."

> **Box 2-7**
>
> ## Codes of Ethics and the Environment
>
> In all engineering codes of ethics the relationships covered are between and among human beings. A most interesting development during the past few years has been the recognition that engineers might also have a moral obligation to the nonhuman environment. Engineering societies, however, have been slow to add such concerns to their codes of ethics.
>
> The first serious attempt to do so was in 1984 by an American engineering founder society (the original engineering societies), the American Society of Civil Engineering (ASCE). A committee suggested the addition of a new canon (the eighth canon) to the existing seven Fundamental Canons of the ASCE Code of Ethics. The suggested canon read:
>
>> Engineers shall perform service in such a manner as to husband the world's resources and the natural and cultured environment for the benefit of present and future generations.
>
> Listed under the canon were nine guidelines to amplify the canon. For example, guideline 8g read:
>
> *(continued)*

Box 2-7 (continued)

Engineers, while giving proper attention to the economic well-being of mankind and the need to provide the responsible human activity, shall be concerned with the preservation of high quality, unique and rare natural systems and natural areas and shall oppose or correct proposed actions which they consider, or which are considered by a reasonable consensus of recognized knowledgeable opinion, to be detrimental to those systems or areas.

But all attempts to have ASCE adopt this canon failed, and the proposal seemed to be effectively dead. However, in 1996 a new approach was taken, and rather than write a new fundamental canon, the first canon was modified to read:

Engineers shall hold paramount the safety, health and welfare of the public and shall strive to comply with the principles of sustainable development in the performance of their professional duties.

The World Commission on Environment and Development (also known as the Bruntland Commission) first popularized the term *sustainable development*, defined as "development that meets the needs of the present without compromising the ability of future generations to meet their own needs" (1987). Sustainable development can be defined in a number of other ways, and indeed the Bruntland Report itself includes ten different definitions. A report for the United Kingdom Department of the Environment contains 13 pages of definitions (Pearce, Markanya, and Barber 1989).

Although the original purpose of this term was to recognize the rights of the developing nations in using their resources, sustainable development has gained a wider meaning and now includes educational needs and cultural activities, as well as health, justice, peace, and security (Herkert, Farrell, and Winebrake 1996). All these are necessary if the global ecosystem is to continue to support the human species. We owe it to future generations, therefore, not to destroy the earth they will occupy. According to the World Bank:

The sustainable approach to development . . . contains a core ethic of intergenerational equity, along with an understanding that future generations are entitled to at least as good a quality of life as the present ones. (Pezzy 1992)

Most engineers would subscribe to this ideal, and on the surface, ASCE has taken a giant step toward incorporating environmental quality into the code. But engineers deal in operations—they are doing things—and therefore need an operational definition of sustainable development. ASCE's definition (in a footnote in the Code of Ethics) defines sustainable development as "the challenge of meeting needs for natural resources, industrial products, energy, food, transportation, shelter and waste management while conserving and protecting environmental quality and the natural resource base essential for future development" (ASCE 1996). This is an amplification of the original Bruntland Comission definition, and still leaves engineers at a loss as to how they are to do this. Another problem with the ASCE sustainable development clause is the wording. The canon states that the engineer *shall* (that's a good start) *strive* (meaning that the engineer has to try, not actually do) *to comply with the principles of sustainable development*. The "strive" undoes the "shall" and makes compliance voluntary. Because there is no operational definition, civil engineers wishing to practice in concordance with their society's Code of Ethics are apparently free to determine what in their opinion are principles of sustainable development, and then the Code asks them only to strive to be in line with what they themselves determine to be these principles.

To its credit, ASCE has convened a series of workshops and conferences on *sustainability* (as opposed to sustainable development) (ASCE 2001). It appears that sustainability is another word for *green engineering*, a concept developed by the U.S. EPA's Office of Pollution Prevention. The green engineering philosophy operationalizes the principles and makes concrete suggestion as to how to solve engineering problems with the minimum of environmental disruption and damage. For example, green engineering emphasizes pollution prevention instead of pollution treatment; looks at the fate of products and analyzes their effects from manufacturing to disposal—the concept of *life cycle analysis;* and generally strives to create minimum environmental impact.

It appears that ASCE is now interpreting *sustainable development* as being synonymous with *green engineering*, and this will go a long way in eventually developing useful guidelines for how engineering should be done. Similar movements have been evolving in other countries as well. For example, in New Zealand the Institute of Professional Engineers and the Australian Institution of Engineers have both developed quite detailed policies on sustainability. In New Zealand this is called *sustainable management* and the policies are those of green engineering.

There is no doubt that the engineering profession has taken sustainability seriously. But trying to not waste any more resources than necessary often leaves the engineer

(*continued*)

> **Box 2-7** *(continued)*
>
> in a quandary because it is not the engineer who decides what project to build. If a client wants a bridge made out of timber, the engineer designs one that uses the least materials. This is both green engineering and cost effective engineering. But suppose that the client wants to cut down a historic hemlock grove in order to build the bridge. The engineer can argue that the hemlock grove has intrinsic value and there are other sources of wood for the bridge, but in the end it is the client who dictates what material is to be used. Will the engineer then not be operating within the principles of sustainable development if she cuts down the ancient grove of trees?
>
> Similarly, Kelly's problem is not answered by going to the ASCE Code of Ethics or any of the related publications. She is going to kill a bunch of fish. That's all. Eventually the fish will return to the river and there is no lasting effect on future generations. Future generations would never know that at one time chromium VI was discharged into this river. And yet Kelly recognizes that this is not the right thing to do. Perhaps, then, Kelly was right after all. Perhaps there is nothing in the ASCE Code of Ethics that would be useful to her. There are no guidelines about what to do about, in Arthur's words, "a bunch of plants and fish."

You think about reminding Kelly that in engineering school she was always very outspoken about ethics and professionalism. She always argued that engineers have a special responsibility to speak up on safety and environmental issues. As president of the local student chapter of ASCE, she'd pushed through resolutions urging ASCE to adopt the controversial proposed environmental Eighth Canon.

But it sounds as if she's already made her decision, and who are you to make her feel bad about it? You remember that you cut a few corners in your first job to please your boss: Where would you be if you'd tried to be a hero? Certainly not on track to be chief design engineer at Pines Engineering Design!

"Well, lot's of luck," you say. "I'm sure you'll make the decision that's right for you."

It's time you left—you have work to catch up on—so you say goodbye, but not before making a tentative arrangement for Kelly to attend your firm's cocktail party next Thursday. Your boss, Joe, is big on engineering cooperation and socializing, and he always wants to meet new people. Such "schmoozing" has more than once resulted in new jobs or new hires for the firm. And besides, Kelly would enjoy meeting your friends at Pines.

As you drive home, you think about Kelly's problem and wonder what you would do in her situation. You're certainly glad you don't have to make those kinds of decisions! Should there not be models, like engineering models, for making ethical decisions, you wonder.

> **Box 2-8**
>
> ## *Ethically Right for Me? I*
>
> Can there be a decision that's "right for Kelly"? Or is there always an objectively right decision? Ethicists are divided on this question. At one level, this division is a disagreement about the very nature of ethics. The tradition that includes figures such as Plato (ca. 428–347 B.C.), Thomas Aquinas (1224–1274), Immanuel Kant (1724–1804), John Stuart Mill (1806–1873), and G. E. Moore (1873–1958) considers that there is always a right answer that the rational person will accept. At the other extreme, subjectivists and existentialist philosophers such as Jean-Paul Sartre (1905–1980) and Albert Camus (1913–1960) and postmodernist thinkers such as Jacques Derrida (b. 1930) believe that there cannot be a "right answer" in this sense. For the existentialists, at least, the nearest thing to a "right answer" is the one that is right for you, given the kind of person you are and your role in the situation. More on this later.

Box 2-9

Ethical Theories as Decision-Making Models

An ethical theory is a high-level account of how its advocate thinks questions about morality should be answered. Ethical theories are in some ways similar to engineering models, and in other ways different. Consider, for example, $F = ma$. This model relates force (F) to mass (m) and acceleration (a). If any two are known, the third can be calculated. This model is a means of predicting what will happen in the physical world. If mass m is acted on by a force F, the acceleration can be predicted. $F = ma$ is a description of how the world works, and its truth is demonstrated millions of times a day. Every single engineer around the world knows that $F = ma$. By applying this equation correctly, the engineer solves—provides right answers to—many engineering problems.

But $F = ma$ is not a law of nature. Albert Einstein showed that $F = ma$ does not apply in deep space, and that given certain circumstances, mass and energy are interchangeable. Nevertheless, on earth Isaac Newton is still right because $F = ma$ works. That is, it is a useful tool for solving many engineering problems.

Similarly, ethical theories do not describe the world. They are prescriptive, and often they provide good answers to moral dilemmas. Though an ethical theory can indeed be used to make a decision (as we explain below), the application of ethical theories is often not as straightforward as the application of engineering models. This is because engineering deals with problems that are clear and, within the framework of the design, can be broken down into separate problems, each of which has a technically correct solution. But people and societies are much more unpredictable, irrational, complex, and "messy" than structures and machines, which work to order and do not have hopes, fears, goals, or ideas of their own. Thus, even people who subscribe to the same ethical theory can disagree about what to do. We discuss this further in Box 4-1.

Moreover, there are many ethical theories to choose from when confronted with an ethical problem, and the selection of the model (theory) will often dictate the outcome. This can be quite disconcerting to engineers who are used to having one model that is acknowledged to be "the truth" so that every engineer will get essentially the same answer if the model is applied to a certain problem. Ethical considerations, in fact, make the practice of engineering more complicated. Qua technical expert, the engineer operates in what lawyer-philosopher Richard Wasserstrom (1975) calls "a morally simplified universe." Wasserstrom thinks that lawyers are encouraged to see their sole obligation to be to the interests of their clients, while staying within the rules of professional practice. Lawyers are required by these rules to take any client who walks into their office, provided they have the expertise and time to do so (and provided the client, once accepted, continues to pay the fees they owe). Lawyers are required not to judge the merits of a client's case, except that they should discourage clients from pursuing unwinnable causes. Engineers, similarly, often see themselves as neutral experts, who ought not to be making judgements about anything except the right answer to the client's problem (though there is no obligation on engineers to work for any client). Once ethical aspects are brought in, though, the situation becomes more complicated, as it does for Kelly.

A third point is that, in practice, most people do not make their decisions by consciously appealing to ethical theories. Nor do they act and decide as if they subscribed to just one theory. Rather, they think that consequences and justice and doing the right thing and taking account of the situation as advocated in the theories we discuss below are all important, and they weigh them up differently in different circumstances.

Nevertheless, a knowledge of ethical decision-making tools can be useful in trying to understand what, all things considered, would the be right thing to do. This is why professional ethics education always includes material on theories. Any theory draws attention to important features of a situation that might otherwise be overlooked. If we subscribe to an ethical theory, we are more likely to be consistent and thoughtful in our decisions, and to be able to justify them to ourselves and others.

In the Western tradition, ethical theories are of two main types: consequentialist (or teleological) and deontological. Consequentialist theories evaluate acts,

(continued)

Box 2-9 (continued)

policies, practices and institutions according to their consequences. Stated simply, in such theories a right action is one that overall has good consequences; a wrong action is one that overall has bad consequences.

The most influential consequentialist theory is *utilitarianism*, developed in great part by English legal theorist and philosopher Jeremy Bentham (1748–1832). His view, often referred to as *classical utilitarianism*, is that only happiness is good in itself. Everything else is good only as a means to happiness. Thus exercise, disagreeable though it may be, is good because it is a means to health, and health is good because healthy people are happier than unhealthy people.

For classical utilitarians, an act is right in proportion as it tends to increase the sum of human happiness or decrease unhappiness and is wrong if it tends to the reverse. Some philosophers have tried to extend utilitarianism to animals and to add the happiness or unhappiness of animals to the utilitarian calculation. This brings up the problem of just how much happiness we are willing to grant to a dog, or a cat, or a pig. More on that later.

Modern utilitarians usually refer to preferences and/or interests rather than happiness, but this is not really a significant difference, because the classical utilitarians believed that happiness is in everyone's interests, that everyone desires to be happy, and that happiness either consists in or follows from the maximum of preference satisfaction. Most current decision strategies (game theory, cost-benefit analysis, risk-benefit analysis, and other derivatives of operations research) trace their origin to utilitarianism.

Other consequentialist ethical theories locate the "good" elsewhere: for example, *ethical egoism* (the good is that which benefits me), *nationalism* (the good is that which advances the state), or *altruism* (the good is that which benefits others).

By contrast, *deontological* theories are nonconsequentialist; deontologists deny that the rightness or wrongness of acts or rules is reducible to the value of their consequences. Deontological theories hold that certain actions are right or wrong regardless of their consequences. The Ten Commandments, for example, represents a well-known deontological system.

The most influential deontological thinker was German philosopher Immanuel Kant (1724–1804).

Kant emphasized the absolute value of persons, who as free, rational beings must always be treated as ends in themselves. To act morally is to follow universal moral principles, which require respect for persons. Kant argued that a test of the rationality of a moral principle is whether or not it can be *universalized*. In his view, rational persons would not be prepared to adopt a rule for themselves unless they are prepared to accept it as applicable to and by *all* persons.

Kant's theory of the universalizability of ethical judgments is but one version of what he calls the *categorical imperative*. That is, some acts such as truth telling are categorically imperative for all persons, regardless of the situation or the consequences. Because these acts are always required, they can be universalized, or expected of all persons at all times. If you choose to act in a certain way, you must then also agree that you must allow others to act in a similar manner.

Kant argues that certain acts such as lying are always wrong, even if you have to lie in order to save the life of another person. He believes that a right action is one done out of "good will" or a respect for the moral law. The good person does right because it is right, and not for any other reason.

The dominant deontological theory in the United States and most other Western countries for over two centuries has been the theory of *rights*. So entrenched is this view of ethics that it is frequently taken for granted that this is what ethics is basically about—for instance, that a topic such as abortion must be discussed only in terms of the rights of the fetus ("right to life") and of the pregnant woman ("right to choose"). A utilitarian approach to abortion, by contrast, would consider the interests of all people involved in the decision.

The English philosopher John Locke (1632–1704) developed a comprehensive theory based on the then-novel notion of natural rights. To Locke, all humans are born with inalienable rights, and these cannot be taken away from us. The most important of these rights, according to Locke, are the rights to life, liberty, and property. Unlike conventional or legal rights, which are bestowed on people by other people, natural rights exist regardless of social acceptance. Locke believes that natural rights are bestowed by God—a

(continued)

Box 2-9 (continued)

view shared by the framers of the United States Constitution—and modern secular claims of human rights are firmly in the Lockean tradition.

Ethics, especially in a legal and political context, particularly emphasizes the mutual recognition and respect of natural rights. However, the security of these rights depends on organized society and government. Ethics therefore requires a "social contract" to protect natural rights. According to Locke, we exchange our "natural" state of anarchy for the guaranteed liberty and security of society. In Locke's view, a rational being will recognize that living in an ethical society is far better than not.

Locke's general point (if not his understanding of history) is valid. Imagine how cruel and miserable our lives would be if we could not count on the honesty and goodwill of our neighbors. Because we all benefit from this social contract, it is our ethical responsibility to uphold it.

Even in Locke's time, the notion of a literal, historical social contract that marked our emergence from a supposed precivilized state was not widely accepted. But the moral ideal behind the social contract—the notion that social organization should rely on mutual agreement and not force—is a powerful idea in ethics. Modern philosophers have used it in the formulation of new theories of ethics. For instance, John Rawls (1971) has developed a theory of the just society that emphasizes respect, impartiality, rationality, and equality. Rawls asks us to imagine a presocial "original position" in which everyone tries to agree on rights and duties before knowing what their position in society will be. Imagine that under this "veil of ignorance," people don't even know their race, nationality, gender, or the period of history when they will live. From this fanciful scenario, Rawls argues that rational people, in forming a social contract, would agree on basic principles of justice: basic liberty for all and distribution of inequalities to the greatest benefit of the least advantaged.

Many philosophers have criticized both the consequentialist and deontological theories as being too restrictive, and have developed alternative approaches to ethics. One such alternative ethical theory is called *situation ethics*, which offers a Christian alternative based on the Ten Commandments. This theory places ethical responsibility on the agent; right actions are those that one does out of love for others.

Existentialists criticize Kant's universalism and focus on the uniqueness of each choice, and on the responsibility of individuals to make their own moral choices. The value of choice, for the existentialist, lies in its courageous assertion of humanity and autonomy in a meaningless universe. Existentialists are fascinated by situations in which conventional morality fails as a guide to action because it offers several conflicting solutions or offers only equally desirable (or undesirable) alternatives.

In recent years, Western ethical theorists have become increasingly interested in *virtue ethics*: the view that ethics is not a matter of bringing about good consequences, nor of carrying out duties, but rather of developing a particular character, of becoming a particular kind of person. Most such accounts are *objectivist*—that is, they claim that everyone ought to aim to develop the same set of virtues, with appropriate allowance being made for different circumstances. Aristotle presented a lengthy list of virtues, each of which he characterized as a mean position between two undesirable extremes. Thus, the courageous person is neither cowardly nor foolhardy, the generous person steers a course between profligacy and stinginess, and the temperate person is neither an ascetic nor a glutton. There will also be virtues specific to particular kinds of people, such as members of a profession.

Aristotle believed that there is a strong connection between virtue and happiness. The virtuous person, in addition to benefiting others, will also necessarily be happier in proportion to his or her own good character. Thus, if everyone were virtuous, everyone would be happy.

The use of ethical models for decision making can be illustrated by the example of Chris's good deed—telling the bartender that he had undercharged. Let's just use a few of the ethical models to see where they lead us.

Take for example, ethical egoism. In this model, one ought to do what is best for oneself. Given that Chris probably would never see the bartender again,

(*continued*)

Box 2-9 (continued)

so there would be no reason to make friends (favors to be collected later), it would have made little sense for Chris to volunteer the undercharge. Just pocketing the money would have been the right thing to do under this ethical model.

On the other hand, using altruism as the model, it would have been clear that Chris, in thinking what is best for the bartender, would have offered to pay the fair bill.

Using the classical utilitarian model, a calculation of cost and benefits is required. First, the people involved would be identified, and their pain and pleasure itemized. The two people involved are Chris and the bartender. What would be the pain or pleasure for Chris to have walked away without paying the bill? Certainly, the money saved would provide pleasure. And there would have been the excitement of having gotten away with something. This positive outcome has to be balanced with the pain or pleasure of the bartender. Suppose he had a family that was in desperate financial shape, and the $58 would have made a big difference in their lives. So now instead of just the bartender, his family is involved, and their pain has to be calculated. Is the benefit realized by Chris smaller or larger than the cost incurred by the bartender and his family? If it is larger, then for Chris to have walked off without paying would have been the right thing to do. If the pain inflicted on the bartender's family outweighed Chris's pleasure, then the right thing to do would have been to pay the correct bill. Of course, we can't quantify pleasure and pain as we can quantity force and mass (though Bentham attempted to do so). But we can use our imaginations. If you were in Chris's situation, you could ask yourself how you would feel if you were the bartender, how your interests (and those of the bartender's family) would be affected by having to make up the difference. The Scottish philosopher David Hume (1711–1776), whose position was broadly utilitarian, believed that this is how we operate in our better moments. Our natural tendency to feel what he called *sympathy* for others leads us to help and avoid harming others.

We could illustrate how many other ethical models can be applied to ethical decision making. The most important conclusion is that the selection of the model often dictates the outcome, and often the outcomes are different depending on the models chosen. Remember, though, that most people do not operate as single-theory appliers. Most utilitarians would not approve of Chris's cheating the bartender even if the bartender did not need the money. One way to accommodate cases like this is to argue that the highest level of utility will be attained if we apply the theory to develop rules that experience will show will overall maximize utility—such as, don't cheat, don't exploit others' vulnerability. Consistently following such rules will lead to greater justice and is indeed the foundation of the social contract. Thus, different ethical theories do not necessarily compete; rather, they draw our attention to different ethical aspects of a situation.

Discussion Questions

2-1. Write a statement for an engineering code of ethics that would be useful to Kelly. If she had found such a statement in the ASCE Code of Ethics, how would this have affected her decision?

2-2. If you were Chris, what else could you have asked Kelly to help her clarify her problem and help her make a decision?

2-3. Should the fact that Kelly's mother is ill have anything to do with her decision? Would Kelly's mother's happiness enter into the decision making?

2-4. Is there a "right answer" for Kelly alone, one that might not be the "right answer" for anyone else? Why would this be?

2-5. What does Alpern (see Box 2-2) mean by "courage" and "moral heroes" with regard to engineers? Is there a limit to the courage an engineer is expected to exhibit? At what point would society forgive the engineer for not choosing to be a moral hero? State some concrete examples.

2-6. Suppose Alison, an engineer with civil engineering training, decides to start a dot-com company selling financial advice to investors (a patently un-engineering job). While on vacation in the woods of New Hampshire, Alison happens to see the new log cabin her neighbor Jerry is building for himself and his family. Jerry has no engineering training and is building the cabin as a hobby (he fancies himself a pioneer!). But casually observing the construction, Alison recognizes that the design is faulty. Jerry has not provided any cross-bracing under the roof, nor has he installed lateral supports to keep the walls from being pushed out. At best, the cabin can start to bulge; at worst, it could collapse catastrophically, probably killing or harming the occupants. She passes on her observations to Jerry, who dismisses her concerns. The Code of Ethics states that an engineer, as a knowledgeable person, has a responsibility to prevent a bad situation from occurring. But she is no longer registered as a professional engineer. Does she still have that responsibility to do something, beyond having warned Jerry?

2-7. Jason is a senior engineering student doing a capstone design project. His work involves designing an earthen dam for a client in central Pennsylvania. In discussions with his professor and the client, the client says that he has no intention of obtaining the proper permits for this dam. It is out in the woods, he argues, and nobody would care. But the dam is upstream from a small community, and if it should burst, it could cause substantial damage and possibly loss of life. The professor shrugs off Jason's concerns, saying that they are just helping the client with the dam and are not really in responsible charge. Jason is not even a graduate engineer, much less a registered professional engineer. Does he have a duty to society to do something about what he perceives to be clearly illegal and unethical behavior?

2-8. Why are college professors not professionals, according to the criteria listed in Box 2-1? Should they be, and how might this come about? What needs to be done to upgrade college teaching to the level of a profession? You might want to talk to several professors about this to get their views.

References

Anonymous. 1957. "I Gave Up Ethics to Eat" *Consulting Engineer* v. 21, December, reprinted v. 39, Dec. 1975, pp. 39–41.

Adams, S. 1996. *The Dilbert Principle*. New York: HarperCollins.

Alpern, K. D. 1983. "Engineers as Moral Heroes" in *Beyond Whistleblowing: Defining Engineers' Responsibilities* V. Weil (ed.) Chicago, Center for the Study of Ethics in the Professions, Illinois Institute of Technology, p. 40.

ASCE. 1984a. The Environmental Impact Analysis Research Council of the Technical Council on Research. "A Proposed Eighth Fundamental Canon for the ASCE Code of Ethics." *Journal of Professional Issues in Engineering* 110 (3) (July).

ASCE. 1984b. Minutes of the Professional Activities Committee of ASCE, January.

Herkert, J. R., A. Farrell, and J. Winebrake. 1996. *Technology Choice for Sustainable Development*, IEEE Technology and Society Magazine, 15, (2) (Summer).

Pearce, D., A. Markanya, and E. B. Barber. 1989. "Blueprint for a Green Economy." Report for the UK Department of the Environment, Earthscan Publication, London.

Pezzy, J. 1992. *Sustainable Development Concepts: An Economic Analysis*. World Bank Development Paper No. 2, Washington, DC.

Rawls, J. A. 1971. *A Theory of Justice*. Cambridge, MA: The Belknapp Press of Harvard University.

Wasserstrom, R. 1975. "Lawyers as Professionals: Some Moral Issues" *Human Rights* v. 5, pp. 1–24.

3

Enhance human welfare

Monday, October 7

"Want you to go to Mifflinburg, Pennsylvania," Joe says. "Prepare a report on the structural integrity of a steel fabrication plant there. Basically for a friend of mine. Take Patrick."

Every time you hear your boss speak, you are reminded of former president George H. W. Bush, who also spoke in phrases without subjects.

"I thought they closed all the steel fabrication plants down there."

"Listening to too many Billy Joel songs. Some industry there. Contract for Cardinal Industries."

"I thought they were in Hoboken."

"Yeah, but just bought out an outfit called Youngstown Metal. Own several plants in the Ohio Valley including this metal fabrication plant in Mifflinburg, which Cardinal wants to close down. Old and unsafe."

"So why'd they buy it?"

"Cardinal wanted other parts of Youngstown, but my friend, the Veep of Cardinal, tells me the Mifflinburg plant is quite unsafe and they want to close it down."

"So why don't they just do—close it down I mean?"

"Ah, well. Apparently Youngstown has a contract with the union that as long as the plant is safe to operate, it will not be closed. If they do close it down for other than safety reasons, there will be large severance packages to the workers. But if the plant is no longer safe to operate, they can close it without the severance packages and just lay everyone off. Escape clause lawyers were able to sneak into the contract. My friend at Cardinal is trying to persuade the union that the Mifflinburg plant is unsafe and needs to close. Too bad, actually. It would mean around a hundred jobs, mostly forty–fifty guys, probably never work again."

You think of your grandfather, who worked in a steel mill all his life, until the mill closed. He spent the rest of his life a bitter man, watching daytime TV.

"Union doesn't know all this, of course," continues Joe. "They see a top-class plant, don't see why it shouldn't stay open and keep the jobs—especially in an area with such high unemployment. They think the plant is structurally safe and are not at all concerned about it. The union believes that if there are safety problems, they can be fixed and that there's no need to close the plant."

"Maybe they can," you suggest.

"Ah, probably would cost much too much, and Cardinal's cash flow problems would not free up this money. Probably needs a lot of work to keep it compliant with OSHA[1] reg-

---

[1] OSHA is Occupational Safety and Health Administration, a United States federal agency that monitors the safety conditions in workplaces.

ulations. Cardinal's line is, they'll have to spend tens of millions, which they'll never get back. But if they close the plant and simply walk away from it, they can at least get something for the real estate."

"So, you want me to do a structural audit?"

"That's it. Go up on Sunday, shouldn't take more than a week."

"Sure. Um . . ."

"Yes?"

"Well, you said this was a favor to your friend, the VP at Cardinal. What if I find that the structure's essentially sound and doesn't need to be closed for safety reasons, or that a minimum amount of fixing can bring it up to OSHA standards?"

"He says it can't be done."

"He's a structural engineer?"

"Of course not. Maybe one of Cardinal's engineers told him? Point is, the union won't necessarily buy that from a company employee, but you being an independent consultant . . ."

There's a pause. "Am I independent?"

"Lord, yes, you have no connection at all to Cardinal."

"You do."

"Yes, yes, remote, remote. Anyway, not telling you what to put in your report. And you won't actually be advising them what to do—that's up to their engineering staff and accountants. Just report what work needs doing, when it needs doing, and what it might cost. Up to the bean counters to figure whether it's worth spending."

Wednesday, October 9

The Youngstown Metal plant in Mifflinburg does not make a favorable impression on you or Patrick. It's located in a decaying industrial area, actually mostly decayed, surrounded by closed-down plants, junk heaps, and abandoned railroad lines. A filthy and apparently disused canal runs along the side of the highway, lined with rotting jetties. An attempt has been made to brighten up the area around the turn-of-the-century buildings with shrubs and trees, and there's a fairly modern looking office building at the front surrounded by flower beds.

You meet the plant manager, who shows you around. The inspection shows that while everything is old, it's been well maintained. You can see no reason why it shouldn't go on for years provided it continues to be taken care of. You meet the plant occupational health and safety and environmental people who are concerned about the plant's ability to comply with OSHA regulations, but it turns out that they had an inspection a few months ago and, except for minor problems, the plant passed. There appear to be no big safety problems, and the structure is old but substantial.

So, the truthful report you want to write won't please management at Cardinal, regardless of what spin you put on it. You will say that the structure is safe, and they would like you to say that it is unsafe.

Your client, however, is Cardinal. How much would you want to please them? And what is "safe" and "unsafe" anyway? If you say the plant is unsafe, it will lead to the plant closing, and drive another nail in the coffin of a dying community. Is this any concern of yours? But the plant is not structurally unsound, even though management would like you to say it is.

> # Box 3-1
>
> ## Moral Responsibilities of Engineers
>
> The first canon of most engineering codes of ethics states: *The engineer shall hold paramount the health, safety and welfare of the public.* The two key words are *shall* and *paramount*. Engineers make a distinction between *should* and *shall* in the writing of engineering specifications. If the word *should* is used, the understanding is that the requirement is left to the discretion of the contractor. For example,
>
> "The contractor should provide a handrail . . ."
>
> is very different from
>
> "The contractor shall provide a handrail . . ."
>
> In the former case, the providing of the handrail is left to the discretion of the contractor. In the latter case, there is no ambiguity. The contractor must provide the handrail, or be in default of the contract.
>
> The second important word is *paramount*. This word likewise is unambiguous. The engineer must place the health, safety, and welfare of the public above all other considerations such as cost and aesthetics. But nothing is ever totally safe, or healthy, or unwaveringly good for people. Thus, even though the requirement is to place public welfare paramount, compromises are made, and the engineer is often placed in a paternalistic role in deciding just how much health, safety, and welfare is required for the money spent. We like to think that there are such people as "the decision makers" who make such crucial decisions, but the fact is that all too often the engineer must decide how much safety to buy.
>
> For example, all highway intersections are monitored to account for traffic accidents and facilities. Based on the number of accidents and the cost of improvements, the engineers decide what modification to make to the intersection. When school boards buy school buses, they do not ask what the survival rate is for children in a crash for each type of bus. They depend on the engineer to design in safety features that are reasonably economical and technically practical. The school board and the parents trust the engineers to design school buses that have the most safety for the cost.
>
> Engineers are required to consider the public welfare because they have special skills on which the public depends, and a cross section of the public could not themselves monitor and evaluate the work of engineering. Because the public, via the state, allows engineers self-regulation, it expects certain benefits from the profession, such as a commitment to honesty, truthfulness, and public service. There is, in short, an implied contract between the profession and the public.
>
> The contract theory of the professions has a striking analogy with the contract theory of ethics first suggested by Thomas Hobbes (1588–1679) and developed by John Locke (Box 2-9). According to some ethicists, the only defensible reason for having ethics is as an agreement of *how we ought to treat each other*. That is, ethics is an attempt to understand and to prescribe a contract among us that would dictate behavior beneficial to all. If you don't lie to me, I won't lie to you, and we will both be better off. Those people who lie (under certain circumstances), we punish by fines or even by sending them to jail.
>
> The same kind of contract exists between the professional and the public. Although it is not directly reciprocal, there is an understanding of behavior that, when adhered to, will benefit all. Part of the contract for professional engineers is that they agree to hold paramount the health, safety, and welfare of the public.

The drive back from Mifflinburg takes you through Amish country—beautiful farms and rolling hills. You pass a few black Amish buggies, red reflective triangle on the back—a compromise solution between the Amish, who do not want anything on their buggies; and the state of Pennsylvania, which insists on having taillights on vehicles. The Amish, you remember, have established a closed society in which people care deeply about each other and shun the outside world as much as possible. Amish men, for example, are exempt from serving in the U.S. armed forces, and their children are required to go only through eighth grade in school. You contemplate whether or not it is morally acceptable for the Amish to live in a free society that tolerates their customs and religions, while not contributing toward maintaining that freedom. Ought it to be their duty to help preserve the freedom they enjoy?

Again, for the umpteenth time, you wonder how it was that you are lucky enough to live in a free society, one that tolerates and even encourages differences of opinion. You wonder what would lead one group of people to become so angry at another group of people as to want to kill them. How is this possible? But more to the point, what if anything is the responsibility of engineers to protect society against such acts of extremism, which we now call *terrorism*? Has the engineer's duty to hold paramount the health, safely, and welfare of the public changed since the events of September 11?

Box 3-2

Engineering and Terrorism

Engineers have not had to think too much about what it is that they are to protect the public health, safety, and welfare *from*. The engineers of bygone days were mostly military engineers, fashioning both offensive weapons and defensive structures. Leonardo daVinci, for example, had hundreds of weapons in his sketch books, and the builders during medieval times built amazingly complex war machines, castles, and other defensive works. If those engineers had thought about their responsibilities to the public, they would presumably have said that they were helping their societies (nations, cities, or communities) both defend against aggressors and enrich themselves by attacking other people. They would have had little difficulty rationalizing the engineering of weapons of human destruction, any more than most engineers today have difficulty arguing that their work in the design and manufacture of armaments is ethical. The *other* that military engineers are protecting the public from is the nation with which their side is at war.

The role of civilian engineers (as opposed to military engineers), who build water and wastewater treatment plants, clean up hazardous waste sites, design highways that reduce fatalities, and build our living environment, has been historically to protect this infrastructure from natural forces, unpredictable accidents, and carelessness/stupidity. But this objective changed on 11 September 2001 with the attack on the United States by terrorists. Now the engineer is required to also protect the health, safety, and welfare of the public from acts of terrorism, and this is quite a challenge. It had never occurred to engineers that they have a responsibility to protect people from deliberate destruction. The two World Trade Center towers were designed to withstand a hit by a Boeing 707, at the time the largest airplane in the skies, and the structures would have survived such a collision had it not been for the extreme heat due to the combustion of the jet fuel. No tall building can survive such a fire, and if this occurrance had been considered when the towers were designed, they may very well have not been built.

Suspension bridges such as the Brooklyn Bridge, the Golden Gate, or the George Washington Bridge can be destroyed by dropping explosives on the vulnerable cable where they are attached to the anchors. Other structures such as tunnels are even more vulnerable. Possibly the most worrisome are our water supplies. Anyone with evil intent can contaminate large sections of the water in our distribution systems and cause massive loss of life before the problem is identified. In short, it is simply not possible to design public facilities so as to make them totally safe from terrorism, no more than it is possible to design anything to make it totally safe from any act of nature (or act of human stupidity).

But the situation is not hopeless. We can begin to understand the problem of terrorism by first asking why someone would do such a deed. This is known in environmental engineering as "going up the pipeline" to solve the problem. Find out first where the stuff is coming from, then fix that, instead of trying to treat the waste at the end of the pipeline.

This is a valid and useful engineering principle applied to the formulation of social policy. Does the responsibility to hold paramount public safety then make it morally obligatory for engineers, who understand such principles, to become more engaged in the political and social processes and to contribute their skills toward defending our freedoms and values?

Thinking of duty brings you back to your job. It occurs to you that there might be such a thing as *corporate terrorism*, but that is too far-fetched. Your duty now is to write the report and submit it to Joe. In the end, your work, suitably edited and technically mangled, might mean the loss of 100 jobs in Mifflinburg. What is your duty here? Is it your job to make sure that this does not happen, or are you a hired hand, doing what you are told?

Box 3-3

Engineers as Intelligent Robots

Many engineers share Chris's concern. Engineering education emphasizes technology since this is the bread and butter of engineering, but non-technical decisions often present the most difficult problems for engineers. For this reason, American engineering education includes courses in the humanities and social sciences and we hope that our graduates have attained a broad education that gives them the ability to think for themselves and to make reasonable decisions that affect society.

Not all engineers receive such an education and sometimes this can result in great social harm. An example of such a case was the changes that occurred in Stalinist USSR. In the 1920s and 30s, Russian engineers were independent thinkers, but such people were deemed dangerous to the Soviet state. Many engineers were executed and most were convicted in show trials of anti-Soviet activity. Eventually the engineers became nothing but robots, relegated to solving technical problems without asking questions. Loren Graham, in his excellent book *The Ghost of the Executed Engineer* (1996) argues that this change in the role of engineering in Soviet society is at least partially responsible for the collapse of the Soviet Union. He recounts an exchange he had in the 1970s with a young engineer.

"What kind of an engineer are you?"

"I am a roller bearing engineer at pulp and paper mills."

"That must mean you studied mechanical engineering."

"No, I studied roller bearings at pulp and paper mills."

The society that educated this young engineer decided that the only thing she was required to know was how to design roller bearings in pulp and paper mills. All other knowledge, technical and social, was considered superfluous (and even dangerous).

Graham describes the events at Chernobyl nuclear power plant prior to the tragic meltdown. Managers in Moscow mandated the tests, and as soon as the engineers started running them, an alarm went off. Instead of alerting the mangers in Moscow, they simply turned off the alarm and continued the test they had been told to run. A second alarm went off, and a third. In all, the engineers turned off six alarms before the meltdown occurred. They did not perceive their job as professionals who should think about their responsibility to society. They were there to perform the test as mandated by the managers, and they were just following orders. The concept of professional ethics was totally foreign to them (Weil 2001).

This is not to suggest that our engineers are like that. We simply want to illustrate an extreme case of poor engineering education. If our engineers are to think for themselves, engineering education has to include courses in global affairs, history of civilizations, and most importantly, professional ethics.

Discussion Questions

3-1. Is there any wiggle-room in the word *paramount*? Would there ever be a situation in which engineers would be required to hold the health, safety, and welfare of the public paramount *unless* . . .? Give examples.

3-2. Do you believe there are any situations in engineering in which the engineer would be morally obligated *not* to place paramount the health, safety, and welfare of the public? Give examples and present arguments.

3-3. The arrangement Chris has is that the report is to be submitted to Joe, the boss, and not directly to Cardinal. Once Chris has done that, it is up to Joe to decide what he sends to Cardinal. Chris believes (but has no proof) that Joe will send a report that Cardinal (the client) wants to receive, a report with which Chris would disagree. Is it at this point any of Chris's business what this report says? If you were Chris, what would you do, if anything? You could use this case to write a short essay on what the limits of public welfare should be for professional engineers.

3-4. In the fight against terrorism, should engineers limit their contribution to developing better technology, or should they become more involved in the political process, using their engineering expertise to help devise public policy? Choose one side of this argument.

3-5. Would you ever kill another human being? If not, why not? If you would, under what circumstances?

3-6. Can you imagine a situation where an engineer could not, concurrently, hold paramount the health, safety, and welfare of the public (all at the same time)?

References

American Society of Civil Engineers. 1996. *Code of Ethics*, Reston, VA.

American Society of Civil Engineers. 2001. *Engineering Forum on Sustainability*, An ASCE/ASEE Newsletter, Reston, VA.

Loren, G. 1996. *The Ghost of the Executed Engineer.* Cambridge, MA: Harvard University Press.

Pezzy, J., A. Markanya, and E. B. Barber. 1989. "Blueprint for a Green Economy," London: Earthscan Publications.

Weil, V. 2001. Illinois Institute of Technology, personal reflection upon returning from a professional ethics workshop in Russia.

World Commission on Environment and Development. 1987. *Our Common Future*, Oxford, UK: Oxford University Press.

4

Hold paramount

Monday, October 14

You're driving to work in your Camaro convertible. You love this car—it's metallic blue and really cool, though Alex despises it and has been known to refer to it as "Chris's grunge-mobile." Alex drives a white supercharged Challenger, allegedly the same model that Barry Newman drove in *Vanishing Point*.

You've had a good week. Pines Engineering Design has landed three big contracts that you've worked on, and you personally sewed up one of them at the cocktail party last Thursday. On the strength of your end-of-year bonus prospects, you're planning to surprise your family with a New Year trip to the Caribbean.

You spend a few minutes at the party talking to Kelly, and you can't resist asking her what she decided about the problem that you discussed over breakfast.

"Oh, I went along with it, you can't always get what you want," she says with a smile. A rather strained smile, you think.

It feels good when you compare your situation with Kelly's. In the five years you've been with Pines, you've never had a serious conflict with management—except maybe the Cardinal report for Joe. You never did see the final report Joe wrote, and you hope it was honest and forthright. He gave you a job to do, and you did it.

Well, you think, that's how it usually is with engineers. Give us the problem and we'll come up with the right answer. This thought reminds you of the presentation you've agreed to give to the advanced engineering design class at De Tocqueville University, your old school, next year. De Tocqueville University is a very prestigious private institution, not quite Ivy League but one of the better private schools in America. It's very difficult to get into DTU.

The talk is a long time away, but you've had a few ideas already. Your topic is "Engineering Excellence," and your theme is going to be that there's always a state-of-the-art answer to any engineering problem. Engineering excellence, you've decided, consists of getting the client to give you enough information so you can understand the problem; setting out the key elements; and finding the most efficient and effective engineering solution. To put it in a nutshell, *engineering excellence is excellent decision making*. And what makes engineering so satisfying is that you're a member of a profession in which everyone agrees to seek the best solution.

Kelly, of course, perceived this as an ethical problem, not an engineering problem. You reflect on how different ethical problems are from engineering. There seems to be room for lots of different points of view in ethics, and not everyone will agree on the best solution to an ethical problem. Too bad the ethicists can't be as efficient as the engineers.

> ## Box 4-1
>
> ## Why Can't Ethicists Be as Efficient as Engineers?
>
> Engineers make technical decisions all the time, from small, on-the-job decisions such as selecting the type of plastic used for a kid's toy, to major, project-level decisions such as the architecture of a new computer or the layout of a major highway intersection. Almost always, they make correct decisions. Engineers share the ability to make technically correct decisions with other professionals such as dentists, doctors, and architects. Lawyers, accountants, and information technology professionals also have to get it right, though the techniques they deal in are legal, financial, and electronic rather than material.
>
> All these professionals are technical experts. Through their training and experience, they develop the professional judgment to decide what is the technically "right answer" to the client's problem, whether it be the construction of a walkway, the removal of a diseased appendix, or the drawing up of a will, to meet the client's requirements. There is always a more or less right solution to any technical problem, at any one time. This is the state-of-the-art answer, the answer on which properly trained and experienced professionals, in full knowledge of the facts (including the client's requirements and the law), will agree. In their endless search for perfection, engineers constantly strive to advance the state of the art, so the right answer in 2010 will be different from what was the right answer in 2000. It will be better.
>
> Throughout this book, we emphasize that many professional decisions have an ethical as well as a technical dimension—consider the examples in Boxes 4-2 and 4-3 below.
>
> When faced with a difficult ethical problem, the professional's first thought is likely to be to consult an expert from the appropriate profession—an ethicist. But you can't expect the same degree of agreement on "the right answer" from ethicists as you would obtain from hydrologists, orthodontists, or accountants. An ethicist will help all parties to a decision to identify relevant issues, to take them into account, and to arrive at a conscientious and good-faith consensus, but an ethicist will not claim to have produced "the right answer."
>
> This is not because ethicists are unprofessional or inexpert. The problem is that there is always more than one ethical perspective, more than one value, more than one set of interests and preferences, and ethicists—and people in general—will attach differing weights to the various ethical considerations. Of course, ethicists strive to achieve consensus, both among themselves and with clients and other stakeholders, and they often succeed. But whereas reasonable surgeons, structural engineers, and lawyers cannot normally disagree about the technicalities of surgery, construction, and legal drafting, reasonable ethicists can and do disagree about the ethical issues we discuss in this book. And so can reasonable professionals and the public.

Traffic is slowing down; you switch on the radio and learn that there's been an accident on the freeway you're using and you're likely to be stuck for a while. Sure enough, traffic comes to a standstill. You switch off the engine and decide to pass the time working on your presentation.

Engineers, it occurs to you, have a special and unique respect for the truth, and that includes knowing their own limitations. Engineers are not jealous of each other's successes; they admire engineering excellence for its own sake and not just as a means to an end. What's more, you think, warming to your theme, engineering really is different from other professions. Doctors are supposed to devote themselves to helping their patients, and certainly lots of them do, but so many of them just want to get rich by selling their services to whoever is prepared to pay top dollar. How can Michael Jackson's and Cher's cosmetic surgeons call themselves healers—not to mention Kevorkian!

Box 4-2

Medical Ethics

The literature on ethical practice in professions such as medicine and law is much more extensive than the engineering ethics literature. This may appear surprising, given that the health and safety of everyone in modern societies is directly determined by engineering whereas most people infrequently need the services of medicine or law. But then, most people never consult engineers, or even imagine themselves consulting an engineer, and do not realize that every time they turn on the light or flush the toilet or drive to work, they are depending on engineering systems for not just their comfort and convenience but their very survival.

In fact, the same ethical issues arise in the practices of all the professions, though there are differences of emphasis. For instance, engineers are required to maintain client confidentiality, but confidentiality issues are much more central to psychology and law than they are to engineering. Privacy issues scarcely arise in engineering practice, but they are an important concern in media ethics. Overall, in developing an understanding of ethical issues in professional engineering, it's useful to study issues from other professions.

In medicine, the first statement in the Hippocratic oath[1] says:

First do no harm.

This rule was meant as a limitation on medical practice: In attempting to benefit their patients, doctors were not to use treatments that might make their patients worse off, nor were they to use their medical skills to cause harm intentionally (for instance, to administer fatal drugs to convicted criminals). In Hippocrates' day (around 2500 years ago), to benefit patients meant to prolong their lives and reduce disability and suffering, while to harm meant to threaten their lives and cause disability and suffering. Given the inadequacy of medicine to provide much in the way of benefits until the mid-twentieth century, these goals were consistent. But, thanks to advances in medical science and technology (antibiotics, life support systems, organ transplants, etc.), these two requirements sometimes conflict, and to prolong life means that suffering is also prolonged. How is a physician to balance these two requirements?

Dr. Jack Kevorkian, a retired physician with a specialty in medical pathology, decided that he would serve the public best by reducing suffering, even if this involved a premature death. Kevorkian did not actually kill his patients nor, apparently, did he solicit patients. Terminally ill people approached him, or were referred to him, with a request that he assist them to end their own lives. He would set up a process whereby they injected themselves with a lethal dose of a drug such as potassium chloride, or in other cases he arranged for clients to be in an airproof tent into which they introduced carbon monoxide, in all cases bringing about rapid and painless death. He was investigated on homicide charges several times and was brought to trial on three occasions, but in each case he was acquitted. Finally, in 1998, he was convicted in Michigan of assisted suicide and sentenced to 10 to 25 years in prison. At the time of writing, he is an inmate at the Kinechoe Correctional Facility in Michigan. He has announced that when released he will no longer assist in suicides but will instead work for changes in the law—physician-assisted suicide (PAS) is illegal at the time of writing in all states except Oregon, and the use of federal funds for PAS was forbidden in 1997.

The voluntary, self-inflicted, and unassisted death of a human being is called *suicide*. When the death is aided and abetted by others such as physicians, it is called *euthanasia*, and this is a major controversy in modern medical practice. Ethicists often distinguish between *active euthanasia*—killing a patient, for instance by a lethal injection—and *passive euthanasia*—refraining from providing treatment that could save the life of a patient who would otherwise die. Passive euthanasia is routinely practiced in hospitals around the world, typically in cases where the patient is terminally ill and where the procedures available would at best prolong life for a short period without improving the patient's already low quality of life. Such procedures are often referred to as "medically futile," because they are of no benefit to the patient.

Passive euthanasia is also sometimes practiced in the case of patients who have a very low quality of life

(continued)

---

[1] At one time, newly minted physicians used to chant the oath in chorus, at graduation. They don't any longer, in most medical schools, because much of it is outdated. For instance, it forbids doctors to "use the knife" (carry out surgery) and to "administer poison" (prescribe drugs, since most drugs are toxic). Nonetheless, its fundamental ethical principles survive intact.

Box 4-2 (continued)

and no reasonable prospect of recovery, but who are not in imminent danger of death, such as patients in a persistent vegetative state. These patients, the best known of whom was Karen Quinlan, are probably unaware of their surroundings and unable to feel pain, though this is not known for certain, and their lives are maintained only by ventilation (heart and lung functions are maintained by connection to an apparatus) and intubation (they are fed by means of a drip). In the case of persistent-vegetative-state patients, passive euthanasia consists of removing these life support systems, at which point the patient usually dies within minutes.

Active euthanasia, which is much less commonly practiced, was legal in the Northern Territory of Australia from 1995 through 1997: Four patients were assisted to die by physicians at their own request. However, the federal legislature repealed it in 1997. Active euthanasia, while technically illegal until early 2002, is quite widely practiced in Holland, subject to strict regulation. In Oregon, the Death with Dignity Act was passed into law through the state's referendum mechanism in November 1994, legalizing voluntary physician-assisted active euthanasia, but due to court challenges it was not immediately brought into practice. In June 1997, the state legislature voted to refer the matter back to the people, voiding the law pending a further referendum. The referendum in November 1997 supported physician-assisted suicide, and in the four years between 1998 and 2001, 140 prescriptions were written, though less than half the patients used the prescribed barbiturates to assist their suicide. This represents less than one death per thousand in the state. At the time of writing, there are a number of legal and political challenges to the law, as well as initiatives to introduce laws similar to Oregon's in other states. Arguments for and against physician-assisted suicide can be found, respectively, at these web sites: www.finalexit.org, www.rights.org/~deathnet/ergo/html, and www.notdeadyet.org.

Dr. Kevorkian's practice did not technically fall under either active or passive euthanasia, because, as we have explained, patients themselves administered the lethal dose that he prepared for them. His supporters believe that he provided a much-needed service to patients who considered that life was no longer worth living and therefore decided to die. It is very difficult to arrange one's own death: Many genuine suicide attempts fail, often resulting in major physical impairment and an even lower quality of life. In this view, if it is legal to take your own life (as it is in most jurisdictions), then it ought also to be legal for a would-be suicide to enlist the aid of a medical expert, thus ensuring a painless and peaceful death. In physician-assisted suicide, the doctor simply helps people to do what they have a right to do. If it's morally acceptable to do something, why shouldn't it also be morally acceptable to help someone do it?

Opponents of physician-assisted suicide (including the American Medical Association) believe that Kevorkian's actions were tantamount to active euthanasia, and that they went against what they see as the duty of the doctor to heal and never to harm patients. "First do no harm" is paramount, they argue. Opponents also believe that allowing physician-assisted suicide may reduce respect for life and make it easier for us to accept clearly undesirable practices such as compulsory euthanasia of sick old people or disabled babies.

Some engineers may believe that they do not have to worry about these matters, since they are not involved in the practice of suicide and assisted suicide. But biomedical engineers have contributed to the design and operation of the technologies that maintain the lives of very ill people who before the design of intensive medicine would have died. Engineers have also worked on technologies of mass killing, including explosive, chemical, and biological weapons and the Nazi death camps. Engineers can also evaluate their professional responsibilities by studying the ethical issues that other professionals must face, especially when these issues are the subject of widespread public debate.

And as for lawyers—shouldn't they be trying to achieve justice? But that seems to be the last thing they care about—were the lawyers in O. J. Simpson's trial remotely interested in getting at the truth?

Box 4-3

Legal Ethics

In the common-law tradition, which is the basis of the legal system in Britain, the United States, and the British Commonwealth, most court procedures take the form of a contest between opposing parties, known as the adversary system. Each party (in criminal cases, one of the parties is the state or the crown) presents their case in the form of a stylized debate, with the outcome determined by a jury or judge. The parties are usually represented by lawyers. Everyone involved in a case is considered to be an officer of the court, including the lawyers for each party, and is supposed to be motivated by a disinterested and unbiased concern to achieve justice. However, the rules of professional legal societies, such as the American Bar Association and the state bar associations, require that the lawyer devote herself or himself entirely to the promotion of the client's case, while still recognizing that the attorney is an officer of the court. For a classic discussion of these issues, see Monroe H. Freedman, "Professional Responsibility of the Criminal Defense Lawyer: The Three Hardest Questions." Freedman (1966) argues that it is not possible to be simultaneously a pursuer of the truth and an advocate for one party to a case. Thus, the defense lawyer (and presumably the prosecutor, though Freedman doesn't discuss this) is always in a potential dilemma (see Box 5-3).

Kenneth Mann's book *Defending White Collar Crime* (1985) presents an interesting study of New York lawyers who specialize in defending clients who are accused of tax violations. The research on which the book is based includes extensive interviews with the lawyers, and the researcher also sat in on meetings between the lawyers and their clients. The author emphasizes that lawyers are legally forbidden from presenting any evidence on behalf of their clients that they believe to be false, because they are not allowed to commit perjury or to assist their clients to perjure themselves. They are also expected to accept whatever information the client presents to them, unless they know that the client is lying. However, as we have noted, their professional duty is to promote the client's interest. Thus, according to the author, the lawyers went to great lengths to avoid their clients' making damaging admissions to them that could hamper the defense case. For instance, in one case the client was accused of fraudulently claiming as a business deduction a family vacation in the Caribbean. The client had indeed had a short business meeting with an associate, but the Internal Revenue Service's position was that this was just a pretext to justify claiming a tax deduction for the cost of the family holiday. At one stage in the consultation, it appeared that the client was about to make a statement to the lawyer that would amount to an admission of attempted fraud: If the lawyer were aware of such an admission, he would be forbidden from presenting the defense that the claim was a legitimate business expense. To protect the client's interest, the lawyer literally jumped out of his chair and clamped his hand over the client's mouth to prevent him from making the damaging admission. Mann concludes that "far from being engaged in the disinterested pursuit of justice, the lawyers were exclusively concerned with winning, where winning is defined as minimizing sanctions for their clients." Clearly, these lawyers see their professional practice solely in terms of providing a service to clients.

But the majority of lawyers would argue that they are indeed in the pursuit of justice for all, and that their work is central to the functioning of a civilized society. They argue that this can be achieved by working for the best interests of each client, resulting in the best and fairest solution to all conflicts. Lawyers believe deeply that the preservation of "the law" is always in the public interest, and they will admit that the law is about process, and not about justice.

In the Rules of Professional Conduct, which govern the actions of all lawyers in the United States, Rule 1.6 relates to confidentiality of information, and the lawyer is beholden to be only the advocate of the client. This requirement can lead to some ethical dilemmas. Suppose a lawyer successfully defends a client who the lawyer knows to be guilty, and the person goes back into society and harms others. Under the rules of conduct, it is not the responsibility of the lawyer to prevent such harm. The most important principle is that the lawyer has no obligation to society (except to practice within the rules)—only to the client. This fact, more than anything else, distinguishes legal ethics from engineering ethics.

You recognize that there is something special about engineers and the engineering profession. Can you put your finger on it well enough to tell the students?

Box 4-4

Jokes about Engineers

Every profession, by virtue of its special standing in society, is a target for humorists and jokesters. Often, these jokes are funny because they describe a characteristic that hits close to home.

The first profession, theology, has been the brunt of jokes for ages, many of which poke fun at theologians' sometimes false piety. Physicians come under fire for their pomposity and often their hubris, and lawyers are usually the brunt of the cruelest jokes, many of which make fun of their avarice or suggest that the world would be a better place without such a large number of them.

> A man goes into an antique shop and finds a large brass rat and decides to buy it. As he is paying for the rat the shop owner cautions him that the rat seems to have magical powers and that he should be careful with it. The man scoffs at such superstitions, tucks the brass rat under his arm, and walks out of the shop. He has not walked more than a block before he hears a sound behind him, looks around, and sees several rats following him. He speeds up, but the faster he walks, the more rats join the procession. Soon he has hundreds and hundreds of rats following him and he is getting scared. He crosses a bridge and decides to throw the brass rat into the water. The rats immediately follow the brass rat into the river and drown. The man goes back to the shop and says to the shopkeeper, "About that brass rat you just sold me . . ." The shopkeeper immediately says, "No refunds. I told you it had some strange powers." "Oh no, I don't want a refund," replies the man, "but do you have a brass lawyer?"

It would be hard to imagine telling a joke like that about engineers, however. As Henry Petroski (2000), an engineer, historian, and observer of engineering culture, points out, engineering jokes seem to poke fun at the proficiency of engineers and are not mean or cruel.

> A lawyer, physician, and engineer are to be beheaded on the guillotine. The lawyer goes first. As the rope is pulled, the blade drops, but stops a few inches from his neck. "A miracle" shouts the crowd, and the lawyer is allowed to go free. The physician goes next, and the same thing happens. As the engineer is led to the guillotine, he looks up at the structure and says, "I think I see the problem."

Engineers are people who see the world as a series of problems to be understood and solved. In the great majority of cases, such solutions are for the benefit of humanity, and thus humanity pokes gentle fun at engineers who are different from others.

> There actually are *three* kinds of people in this world. The first kind look at a glass half full of water and say that the glass is half empty. The second kind of people say that the glass is half full. The engineers, however, look at the half empty glass and say that the glass is too big for the volume of water it's meant to hold.

Basking in the warm waters of professional complacency, you're sharply interrupted by the impatient honking of horns, and you realize that the traffic has started to move again.

Monday, October 14

Your upbeat mood evaporates the moment you arrive at work.

"Isn't it awful, the crash?" your secretary Rosemarie asks. You look blankly at her, so she explains that 12 employees of Pines died in an accident last night. They'd been playing in the annual softball tournament between engineering firms in the region, held this year in a mountain city about 100 miles away; on their way home, a truck heading toward them skidded out of control and swiped their bus off the road and over a 1000-foot drop into a lake. All the 27 people on the bus—driver, players, families, friends—were killed.

You are speechless. You go to your office, check your e-mail, and the first message is from Joe, outlining the tragedy and summoning all the senior engineers to an emergency meeting at ten.

Everyone is shocked and very distressed. They've all lost close friends in the accident, including two highly respected engineers who were each with the company for over ten years. After expressing his sympathy for the victims and their families, and assuring the meeting that the company's insurance will protect the employees' families financially, Joe hands over the meeting to Sarah, the chief design engineer.

Sarah is a formidable woman in her 60s who is widely respected. She was one of the first women to attend Purdue engineering school, and she really had to struggle hard to be taken seriously. In the process of achieving her success, she developed a certain ruthlessness, and behind her back is referred to as "The Iron Horse." On this occasion, however, she is visibly upset.

"I know we're all shattered by this tragedy," she begins, looking around at the tear-stained faces. "But life has to go on, including meeting our current contracts. I'm sure our clients will be sympathetic if we ask for extensions on deadlines, but if we can do it, I think we owe it to our friends and colleagues to finish off the jobs they were working on just as they would have done—on time. We can hire temporary help—the president of one of our biggest competitors already called to offer to loan us a couple of good people—but we're all going to have to take on extra work. It'll mean working some weekends and late nights for a while, and some of you will be spending time on fairly basic design work that you haven't done for a while."

One of your colleagues says quietly, "Whatever it takes!" and there's a murmur of agreement.

The rest of the day is spent in reorganizing project teams and allocating jobs. By the evening everything is organized, temporary help has been arranged, and everyone goes home.

Box 4-5

Engineers Working Together

Leaders within the engineering profession often lament how invisible the profession is. There are no famous engineers. Architects, physicians, lawyers, scientists, and many others get press coverage and appear on the cover of *Time* magazine. When was the last time an engineer was on the cover of *Time*?

There are two reasons for this lack of public glamour. First, the public expects engineers to succeed. A newsworthy event would be if something did not succeed—a bridge collapsed or a computer crashed. A bridge that functions as designed will seldom be in the news, and the engineer who designed it will be totally invisible.

The second reason engineers are not public figures is most engineers do not seek publicity and find that they are best rewarded by solving the technical problems. Engineering is also well compensated, of course, but there appears to be little downright avarice among engineers. Creating something that did not before exist so that people can use it is the greatest reward.

In today's complex world, to achieve any notable engineering success requires a team effort. Gone are the days of the single engineer conceiving, designing, and building a gadget that has great impact on the public welfare. Engineering is a *social* enterprise, quite unlike medicine or law where the professional deals mainly with a single individual. In engineering, it is the *problem* that is important and that needs a solution. This leads engineers to cooperate rather than compete. The fun of engineering is watching the thing get done, whatever that thing is, and the satisfaction comes from contributing to its completion.

Tuesday, October 15

One of the projects you inherit is a $50 million, 700-room, 29-story, 320-foot-tall hotel—the Asmara Philadelphia. Pines is designing it for Timmo Holdings, a national property company. The Asmara, which is well under construction and is due to be commissioned in six months time, looks like a straightforward building, but you know from experience that there are always problems with a large project.

Over the last few years, your job has steadily become more managerial and supervisory—endless meetings!—with correspondingly less of the hands-on, problem-solving work that attracted you to engineering in the first place. So, despite the tragic circumstances, you're looking forward to putting on a hard hat again.

You've taken to heart Sarah's statement that you owe it to your deceased colleagues—and to Pines—to finish the job for them. You're determined that the Asmara Philadelphia will be a triumph of engineering design and a fitting memorial to Ali, Dianne, and Jerry who worked on the design—they all died in the wreck.

You go home and spend the rest of the day—and most of the night—checking out the design. Around 10 P.M. your antennae suddenly begin to twitch, just as Alex walks into your study with a pot of coffee.

"What's wrong with it?" Alex, antennae observer, asks.

"Mostly it's fine, basically because it's pretty bog standard construction, but the entrance . . ."

"*What* standard?"

"Oh, bog standard. A bog means a toilet in England. 'Bog standard' means conventional boring design like a conventional public restroom, and . . ."

"Like 'plug and chug'?"

"Yes. But the overhang on the entrance, it's real complex, and I have a funny feeling about it."

"Looks nice to me," Alex says. "Like the sails on the Sydney Opera House. Cool feature on a boring building."

"Didn't we have a great time in Sydney? But the Opera House, it had cost overruns Kevin Costner would have been ashamed of, and I think these shells were just dashed off in no time at all. I've done a lot of work on structures like this, believe me, they're not that easy. This one looks underdesigned to me. I'm trying to imagine you and the kids standing under it on a really windy day, and I don't like it."

"Do you know how to fix it up?"

"Oh sure. Tear it down, design it properly, rebuild it."

"It's already built?"

"Finished last week."

"Oh."

Box 4-6

A Technical Challenge

The overhang at the hotel entrance that Chris is worried about is to be cantilevered out from the wall, with sculptured thin interlocking shells—resembling the Sydney Opera House's "sails."

Such a design is difficult and requires advanced methods of engineering analysis. One such method is *finite element analysis*, in which a structural component is divided into many smaller individual interconnecting elements, each interacting with each other. By studying the behavior of these elements, the overall loading on the structure can be analyzed and the properly sized members designed.

(continued)

Box 4-6 (continued)

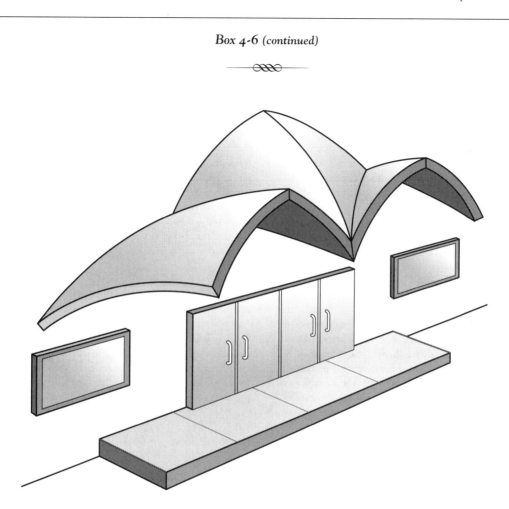

For example, a finite element analysis of a simply supported beam might be done by first dividing the beam into a number of elements, as shown below:

Each finite element is analyzed independently, recognizing that each element interacts with each other element. The more elements there are, the greater is the number of interactions. For this reason, finite element analysis is highly computer intensive. A complex structure such as the one designed by the architects for the Asmara Philadelphia requires a long and time-consuming analysis. Running such computer

(*continued*)

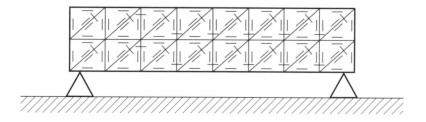

Box 4-6 (continued)

programs can take days, a luxury that the design team evidently did not have.

Analyzing the structure with a fewer number of finite elements can reduce computer time. That is, instead of assuming several thousand small elements, all interacting with each other, the engineer can assume only a few and dramatically reduce the computer time. For example, the beam can also be analyzed with the elements as shown below.

If the structural member is a simple one and the way it is placed is also not complicated, such simplification is permissible. An engineer can in many cases sense from experience if a structural member is correct or not. Shells such as the ones designed for the hotel, however, are extremely complex, and a reduction of elements for such structures can lead to gross errors. In addition, shells are notoriously difficult to analyze at the locations where they intersect. Intersections, such as the ones designed for the hotel, require thorough and detailed analysis.

Another problem faced by engineers is predicting the loading on the structure. The mass of the structure itself is called the *dead load*, and other loads are called *live loads*. Live loads include such forces as wind, snow, and even people crawling on the structure.

Once the loads have been established and the structural analysis has been done, concrete structures (such as the overhang) are designed by calculating the number and size of the reinforcing bars and the concrete thickness necessary to hold certain loads. Always called "re-bars" in engineering talk, reinforcing bars are the steel bars placed in concrete to give it tensile strength, or strength against being pulled apart. The placement and number of re-bars is critical in the ability of concrete to withstand loads. If the analysis is wrong—that is, incorrect stresses in the structure are predicted—then the design, such as the placement, size, and number of re-bars selected, will be wrong, and the structure may not be able to carry the actual loads placed on it.

In our scenario, the engineer in charge was pressed for time and decided to do a simplified finite element analysis, dividing each steel member in the shell into one row of elements (like the simplification for the beam below). This greatly simplified the calculations and allowed him to finish the job on time.

The design, based on the crude analysis, was faulty, but not obviously faulty. No engineer could detect the problem just by looking at it because it was a complex series of arches. The problem became evident only when Chris, an experienced structural engineer, thought it "looked inadequate," and then ran his own analysis of the shells.

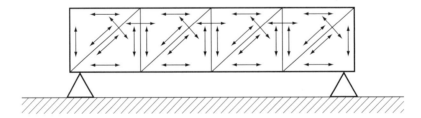

Wednesday, October 16

You arrange a meeting with Sarah and Joe to discuss the Asmara project and explain your concern.

"You can't say who did what because they worked as a team, but it looks like Jerry did the structural design for the overhang. He was a ME,[2] wasn't he? Had he even done a concrete design before? And these aren't basic design errors, so the others probably wouldn't have picked them up. But Ali was supposed to be the project leader and he should have checked everything before sealing the drawings."

[2] mechanical engineer

Box 4-7

Engineering Qualifications

Society yields to all professions powers to do certain things that would be illegal for nonprofessionals. For example, the physician is allowed to cut people up. If anyone else did it, they would end up in jail. Similarly, police are allowed to shoot people. Engineers are allowed to seal drawings. That is, an engineer's seal is required on all drawings that are for public facilities or that are constructed with public money. Only licensed professional engineers are allowed to seal drawings, and when they do, they stake their professional reputation on the accuracy of those drawings. If a facility fails, the engineer cannot blame anyone else. The buck stops with the engineer who sealed the drawings.

"You are very sure this is a serious problem?" asks Sarah, "We are not looking at a trivial matter? What do you estimate is the factor of safety for the overhang, as it is now?"

"Jerry expected the wind load to govern design, and I agree," you respond, "and for a 100-year design wind, the factor of safety is less than one."

Box 4-8

Factors of Safety

Engineers all use factors of safety in the design of structures, processes, materials, or any other product of engineering. For example, a submarine is designed to withstand a certain maximum water pressure, such as 1000 feet of water. But because of unknown factors such as quality of welds, variability in the strength of the steel, and so on, engineers might actually design the submarine to withstand pressures from 1500 feet of water. This yields a *factor of safety* of 1500/1000 = 1.5. This way the design engineers are quite confident that the submarine will not be crushed with a pressure of 1000 feet, and quite conceivably, might be able to survive 1500 feet, but the engineers would not themselves dive to such a depth to test it.

Structures, such as a building, commonly have large factors of safety because it is impossible to predict what a building will eventually be used for. A building may be designed as an office building, but 50 years later it might become a storehouse for books, resulting in a much higher floor loading (pounds per square foot of floor). For commercial structures, therefore, a factor of safety of 2.0 is not unusual.

(continued)

> **Box 4-8** *(continued)*
>
>
>
> In some cases, such as airframes for airplanes, the factors of safety are much lower since the cost of operating an airplane increases by a factor of four for every extra pound that has to be lifted. Because structural components of the airframe are carefully designed and all excess weight is eliminated, a factor of safety of 1.2 is typically used.
>
> The factors of safety are calculated on the basis of the design load. For example, a shear pin on a centrifuge that prevents excessive torque from damaging the motor or gearbox might be set to break at 200 foot-pounds of torque. If the centrifuge is designed to withstand 300 foot-pounds of torque, the factor of safety is 1.5. But suppose the wrong shear pin is put in the machine, one that will break at 400 foot-pounds. In this case, the factor of safety is actually less than one, of 300/400 = 0.75, but it does not mean that the machine will be damaged. If the maximum torque the machine is required to handle is always less than 300 foot-pounds, the centrifuge will probably not fail. If, however, the torque exceeds 300 foot-pounds, and the shear pin does not break, then extensive damage could occur.
>
> The same is true for the Asmara overhang. The design loading might have been a 100-year wind, or a wind that is expected to occur once every hundred years, and the original design might have had a factor of safety of 2.0. What Chris found out was that further analysis might show that the factor of safety is less than one for this design wind, so that theoretically, a wind of greater than say 50 miles per hour, which might occur once every five years, would probably cause it to collapse. If Chris is right, the overhang is quite dangerous. But Chris might not be right, and the actual design might be adequate.

There is silence.

Finally Joe says: "I've already submitted the building for several engineering achievement awards." You fleetingly remember that engineering achievement award nominations seldom come from satisfied clients, who are too busy putting the products of engineering to use to think of how well the products perform. Firms often go to great lengths to publicize their own achievements, and if they win the competitions, their clients are both surprised and greatly pleased that they obviously had the good sense to select an award-winning engineering firm.

More silence.

Finally Joe asks you: "How long do you need and what's it going to cost us?"

You know that Joe appreciates honesty. "I haven't had time to do detailed costings, and I've never worked on a problem like this before, except theoretically, so it's a bit of a guessing game. For sure, we'll have to tear down the overhang, redesign it, and rebuild it. We'll have to negotiate a change order with the contractor who'll know we're in a hole and charge us accordingly. Say five hundred thousand dollars and six weeks, and hope the weather isn't too bad. That'll mean they can move in on June 1."

"Ok," Sarah says, "Joe, can you talk to Timmo's people and arrange to delay the completion date?"

"Half mil. OK," Joe says, wincing. "Don't know how they'll feel about the six weeks, but will try."

Friday, October 18

Over breakfast, you tell Alex that you're feeling distinctly pessimistic. What if Timmo doesn't agree to the extension? Will Joe insist on the redesign or just let the overhang stay as it is?

"The problem is, Joe thinks engineers are wonderful. I had a long conversation with him a year ago at a convention. He told me that he majored in classics at Yale and used to really look down on scientists, especially engineers."

"Joe *looks* like a classics professor," Alex says. "All that wispy white hair and those terrible tweed jackets with the leather elbows. And as for that heap of junk he drives around in . . ."

"He says it's going to be a classic," you say mildly.

"No it isn't, there are no classic Japanese cars. Now it's just old. If it lasts long enough it might get to be very old."

"Well," you continue, "while he was working on his doctorate he met Brenda, who was a ME instructor, and he fell in love with both her and her subject and . . ."

"I didn't know Brenda was an engineer. She runs a landscaping business."

"Anyway, Joe abandoned classics, got a job in a bank, and studied for an MBA part-time. When he and Brenda set up Pines, he took over the company management while she did the technical stuff. Lucky for Pines, he turned out to be a spectacularly good manager."

"But Brenda is no longer with the company?"

"She got bored with engineering not long after she and Joe started Pines and decided to make a change. Joe was devastated. He'd been relying on her to provide the engineering knowledge, but luckily Brenda had worked with Sarah and knew she was looking for something new too, so she introduced Joe to her and it all clicked."

"It must be nice having a boss who appreciates professionalism," Alex says. Alex is dean of business studies at McGregor University, a prestigious private school, whose president—himself an ex-classics professor—thinks commerce is vulgar.

"Yes and no. He likes to think he's one of the guys with the engineers, but he doesn't really understand engineering. He loves stories about brilliant engineering successes.

Box 4-9

Engineering Triumphs

Modern engineering boasts a string of triumphs that have forever changed our lives. Consider a world without just some of these—electric light, airplanes, plastics, computer chips, penicillin, steel, satellites, television, aspirin . . .[3] In fact, we rarely think of these as engineering triumphs. They are just part of our everyday world.

The kinds of engineering feats that our character Joe would most appreciate, however, would be structures. He would be impressed by the pyramids and amazed by the aqueducts in Rome. How could the engineers have achieved such straight lines without having any surveying instruments? He would wander through a city like St. Petersburg in Russia, not admiring the paintings in The Hermitage, but being impressed by the city itself—how Peter the Great chose to build this city in a swamp. The engineers had to

(continued)

---

[3] One of the authors (engineer Aarne) has long held that the three greatest engineering feats of all time are duct tape, bungee cords, and WD-40. But he is painfully practical.

Box 4-9 (continued)

bring in all the stones and boulders, one by one, in order to lay the foundations for the magnificent buildings. Joe would make the pilgrimage to the Brooklyn Bridge in New York, perhaps the greatest single American engineering success of the nineteenth century. He would know all about John Robeling, the original designer of the bridge, and his son, Washington Robeling, and Washington's wife, Emily Robeling, who acted in many ways as the chief engineer after Washington was felled by bends. At the time the bridge was designed, the caissons used for the two towers were the largest by far that had ever been constructed. The superstructure was constructed of steel, the first such steel bridge in the United States. The central span was high enough to allow clipper ships from the Brooklyn Navy Yard to pass underneath. The construction took a full 14 years, from conception to completion, and the celebration at the opening of the bridge was the greatest New York had ever seen. Chester T. Arthur, the president of the United States, was on hand to formally open the bridge, and Emily Robeling had the honor of being the first person to cross the bridge in a carriage. The Robelings and their assistants, men like C. C. Martin, Francis Collingwood, Wilhelm Hildenbrand, E. F. Farrington, and George McNulty, would all go on to successful engineering careers.

While there are many books on the Brooklyn Bridge, we would recommend *The Great Bridge* by David McCullough (1972).

Of course, he knows about the failures too, but he sees them as *managerial* failures. He says that the engineers in cases like the Challenger didn't make technical errors; in fact, they pointed out the problems to management but were ignored."

Box 4-10

Engineering Failures

Engineering produces things that succeed. A bridge, for example, is supposed to carry a certain traffic volume from one shore to another. If it does, it is a success. A motorcycle is supposed to have certain acceleration and attain a specified fuel efficiency. If it does, it is a successful machine. But not all engineering projects end up as success stories. Engineers are human and can make mistakes.

Engineering failures can be categorized into three types:

1. The first kind of failure is as a result of a mistake. The computer age has produced a plethora of such failures, exemplified perhaps most notably by the multimillion-dollar satellite that failed to function because someone had mixed up the metric (SI) and customary American units. This is a human error, and although engineers are very careful to check and recheck work before it leaves the office, some errors occur, but seldom with the same consequences as the ill-fated satellite.

2. The second kind of engineering failure occurs when the structure, or machine, or electrical component, or chemical plant fails as a result of an unanticipated load or condition. For example, suppose a building is designed for a certain wind load (the force of the wind pushing against the building) and the design engineer assumes that the building is symmetrical, so that the direction of the wind does not matter. But this may be a bad assumption. Maybe if the wind comes from a different direction, the forces on the building's structural elements are much higher than anticipated. If the engineer does not recognize this and designs the building using only one wind load, the building might be toppled by a high wind from the "wrong" direction.

(*continued*)

Box 4-10 (continued)

(This is the case of the Citicorp center discussed in Box 4-12). Obviously, if the building falls down, it is an engineering failure.

3. A third kind of engineering failure occurs when, although all the forces are anticipated and the design is correct, there is a side effect of the structure or device that causes unanticipated harm. An example of such a failure was the Pruitt Heights high-rise public housing structures built in East Saint Louis. The buildings, from a technical perspective, were successful, but the lifestyle of the occupants was not compatible with the buildings. After much effort to change the living habits and values of the occupants, the housing authority finally gave up and the buildings were demolished. As buildings—steel and concrete—they were successful. As low-cost housing, they were a disaster.

Other products that have desirable engineering properties can end up being failures. For example, one of the most important advances in electrical engineering was the development of polychlorinated biphenyl (PCB), a fluid used in electrical transformers. The engineers wanted a fluid that did not biodegrade (no microorganisms were interested in using it as a source of food because it was toxic), did not degrade under high temperature, was nonreactive with the transformer walls, had excellent heat transfer capability, and was chemically stable for long periods of time. PCB was the ideal transformer fluid, and this product was used for many years in all large transformers—a clear engineering success.

But then engineers and scientists began to discover the negative side of PCB. Its persistence, when discharged to the environment either by accident or on purpose, was a problem because it is toxic to life (that's why it did not biodegrade). Worse, it bioaccumulated, so that the animals that fed on smaller creatures would accumulate PCB in their bodies. Human beings, who eat high up the food chain, were soon accumulating PCB to levels that were considered dangerous. Eventually, most PCB-filled transformers were emptied and the material destroyed, but there is still PCB around and accidents continue to occur. PCB was an engineering success story until its secondary effects were discovered. Now it is considered an engineering failure.

In summary, there are three kinds of engineering failures:

1. Failures due to errors in calculation or fabrication
2. Failures due to not anticipating all the ways the design can fail
3. Failures that occur because not all of the side effects were predicted

Perhaps there is a fourth kind of engineering failure: when the structure or device is used for immoral purposes and should not have been built at all. Ready examples include the gas oven in the Nazi concentration camps and the production of the gas used to kill the people before burning them. One could argue that this is not an engineering failure, but a human one. But it was engineering, the field that is supposed to hold paramount the health, safety, and welfare of the public, that failed.

"Does he think the engineers were right to let the Challenger launch go ahead, then?"

"No, he thinks they were spineless fools. But the point is, those disasters *did* happen and, he thinks, it was obvious—if not to the poor astronauts—that they would happen, so there was no excuse. But he also thinks that, given time, a top engineer can solve any problem. And he prides himself on hiring only top engineers."

"Like you."

"He thinks so, yes."

Ruefully, you recall the presentation for the engineering class that you've been planning. Your faith in engineering excellence now looks impossibly smug.

Box 4-11

Engineers as Managers

Does an engineer's responsibility to his or her employer ever override the duty to place the health, safety, and welfare of the public paramount?

This question is best illustrated by the sad events of the Challenger space shuttle, which crashed in 1986. The events leading up to the disaster are well known.

On the night preceding the launch of Challenger, the engineering team at Morton Thiokol, the manufacturer of the booster rocket, were asked to approve the launch. They argued against launching because of the cold temperatures and the worry that in such cold temperatures the seals between the rocket sections would not hold.

Unfortunately, they were overridden by the management team, who then recommended the launch.

The decision to launch or not to launch was a management decision, not an engineering decision. Jerald Mason, senior vice president at Morton Thiokol, knew that NASA was looking around for another source of booster rockets, and the loss of this contract would have been devastating to the company. NASA was also under pressure to launch, ostensibly from the White House, where President Reagan was writing his State-of-the-Union speech, set for delivery the following evening. But the information from the engineers was troubling, and the management team had to make a decision.

One of the managers, Robert Lund, supervising engineer (for whom all the engineers worked), was loath to go against the recommendations of his engineers. While the data were inconclusive (the engineers could not guarantee that the O-rings would fail), the probability was quite high, and such a failure would have spelled disaster for the Challenger. But at a crucial time in the deliberations, Jerald Mason turned to Robert Lund and asked him to "take off your engineering hat and put on your management hat." (Report of the Presidential Commission, 1986, p. 772) Thinking like a manager, Lund agreed to go along with the launch.

Following the explosion and the loss of the astronauts, the company and NASA both tried to cover up the cause. It took Roger Boisjoly to break with the company and bring the facts to the commission investigating the disaster. His courage allowed the truth to emerge, and for the real cause to become clear.

The interesting person in this unfortunate event is Robert Lund. On Mason's urging, he was able to take off his engineering hat and put on his management hat, and this allowed him to reverse his original position. At the hearings following the explosion, Mason was asked what he had in mind when he asked Lund to put on his management hat.

"I had in mind the fact we had identified that we could not quantity the ... movement of the primary [O-ring]. We didn't have the data to do that, and therefore it was going to take a judgment, rather than a precision engineering calculation, in order to conclude what we needed to conclude." (Report of the Presidential Commission, 1986, p. 773)

Mason's argument seems to be that since the engineers could not prove without doubt that the Challenger would explode, the decision became a management decision, and that Lund should then have to think like a manager. But Mason, like Lund, also had an engineering background, and ought to have understood that engineers never know anything for sure. Engineers work on the basis of probabilities, and in the case of the Challenger where human lives were at stake, the benefit of the doubt should have been on the side of safety.

The question that Boisjoly and the other engineers faced when they were rebuffed by the management team was what to do next. They could have called the astronauts (or tried to), or they could have gone public. It appears that they did not think of doing either. They were good company men. To blow the whistle (see Box 5-6) would probably have been the end of their careers, no matter what happened next. It would have been a great sacrifice, and if the Challenger had not exploded, it would have been for nothing.

Or would it? They could have acted on the probability of an accident occurring and have been justified

(continued)

> **Box 4-11 (continued)**
>
>
>
> even if it hadn't occurred. Consider this analogy: When the TV weather forecaster tells you that there is a 70% chance of precipitation tomorrow, you do not complain if it doesn't rain after all. You plan your day based on this assessment. It could be argued that the Challenger engineers had a duty to act with professional integrity and announce the risk. It would also be easier to sleep at night, knowing that one had at least made the risk clear. You have discharged your responsibility by assessing and communicating the risk: What people do on the basis of that assessment is then their responsibility.
>
> Diane Vaughan, in her book *The Challenger Launch Decision: Risky Technology, Culture, and Deviance at NASA* (1996), argues that NASA personnel—engineers as well as managers—had gradually become more accepting of minor failures, a phenomenon she calls "the normalization of deviance," which made disastrous failure progressively more likely to occur. Certainly, the engineers at Morton Thiokol, the company that manufactured the solid fuel rocket boosters for the Challenger (and other shuttles) had been concerned for several years about the strength and flexibility of the O-ring sealing mechanism, and indeed NASA itself was concerned about sealing problems as early as 1977. While design changes to the joint were made, if anything they may have increased the risk of failure, and after the 17th successful flight, the NASA manager of the solid fuel rocket booster described O-ring erosion as "accepted and indeed expected—and no longer considered an anomaly" (Werhane 1991; Searbruck and Milliken 1988).

Friday, October 18

Joe calls you and Sarah to another meeting. True to Alex's description, he's wearing a worn Harris Tweed jacket, khaki pants, battered brown oxfords, a plaid wool shirt, and a red bow tie.

"Bad news! Timmo people are really proud of the Asmara. They've arranged an international architects' convention for the opening of the hotel in late May—over a thousand delegates and partners, every room and convention facility booked, huge media coverage."

"Well," you say, "the overhang has to be torn down. It won't be safe unless it's completely redesigned and rebuilt. We can't have the hotel ready until June 1. Can't they change the date?"

"Are you serious? You can't just switch the dates on a major international convention like this. They've got the mayor of Philly and Pete Rose opening the convention, and the guest speaker at the dinner is architect I. M. Pei. Afterwards, there's a charity ball with Barbra Streisand, Garth Brooks, Vanessa-Mae, Gillian Anderson, and Andrew Newton—the whole nine yards, wouldn't be surprised if Buckminster Fuller and Elvis show up!

"They've allowed for a couple of weeks delay, and they'll need four weeks to set up the convention—April 30 is absolutely the final date for them to move in. I suggested they try to arrange an alternative venue, and they just laughed."

"But didn't you tell them about the problem with the overhang?" you ask.

"Come on! How can we admit to designing a building that might fall over? How much work do you think we'd get from Timmo? And bad news travels fast—we'd be ruined!"

Box 4-12

Acceptable Risk

The concept of acceptable risk is perhaps best illustrated by the story of the Citicorp Center in New York, a spectacular high-rise building built in 1972.

Citicorp (First National City Corporation) decided to build an attractive, functional, and imaginative 59-story building for its headquarters. The site, however, had space restrictions due to a small church building on the corner of the lot. The architectural firm, Hugh Stubbins and Associates, struck a deal with the church. The building would be designed with supporting columns not at the corners of the building, but in the middle of each side. This allowed plenty of space to build a new church on the corner of the lot. The four nine-story columns made the building look light and airy, but presented a challenge to the structural engineers.

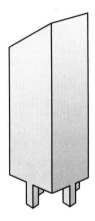

The structural design was to be done by the engineering firm headed by William J. LeMessurier (pronounced "LeMeasure"), one of the leading structural engineers in the world. He decided to use a unique form of construction, with the wind and gravity loads being transferred to the four columns by means of V-shaped or chevron braces, a brilliant solution to the problem.

(*continued*)

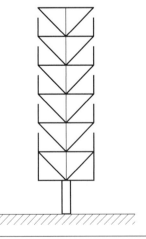

Box 4-12 (continued)

The design of such a unique structure presented unprecedented challenges. Design is an iterative process. A structure is postulated and the loads on that structure are then estimated. Using mathematical principles and well-tested equations, the effect of these loads on the structure are calculated. In the case of the Citicorp building, LeMessurier engineers calculated, in addition to other live loads, the effect of wind and decided that with a damper mechanism in the top level of the structure the building would be able to withstand the design winds and would not sway so much as to make the occupants queasy. The effect of wind is complex in a city with high buildings, but almost always the greater force on a structure are winds hitting the building from the side, and these were the winds that were studied in wind tunnels.

The building was constructed and occupied, and the client was very pleased with the result.

Six years after the building was occupied, LeMessurier got a telephone call from a student who had been assigned to analyze the Citicorp building. The student said that his professor had told him that the columns had been placed at the wrong place. LeMessurier patiently explained the circumstance to the student, and told him that he now knew more than his professor. But this conversation led LeMessurier to think further about the design. He wondered if the chevron wind braces were just as strong if the wind direction was diagonal, so-called quartering winds. To his surprise, the calculations showed a 40% increased strain in the wind braces. This would not have been a problem, except that LeMessurier learned that the wind braces, which were much too long to be a single steel beam, were bolted instead of welded, as they had been in the original design. Such changes are routine in engineering offices and the design had been done according to code. But the engineers who agreed to use bolted connections instead of welded connections did their calculations assuming that the main function of the chevron braces was to carry the gravity load instead of bracing against the wind load. The connections were thus much weaker than they should have been.

Le Messurier was now worried. He went to his summer cabin and redid all the calculations, beam by beam, to see where the weakest part of the building was. He discovered to his horror that the building would be in danger of collapsing if a 16-year quartering wind hit the building. That is, such a wind would be expected statistically to occur once every 16 years—a clearly unacceptable risk. Knowing this, LeMessurier considered his options. He could keep silent and hope for the best; he could kill himself; or he could blow the whistle—on himself! He decided not to kill himself because he was curious as to how the problem would be resolved, and the first option was untenable because, as he said later, "Thank you Lord, for making the problem so sharply defined that there is no choice to make."

LeMessurier went to the owner, and with their cooperation and encouragement, working with the architect, disaster prevention teams, insurance agents, and construction firms, was able to get into the building and start welding all the connections on the wind braces from the inside.

A short time after the work began, hurricane Ella decided to head toward New York City. This was a 200-year storm, and had it hit, not only the Citicorp building but other tall structures might have been destroyed as well. At the last moment, the hurricane veered off to sea. By the end of the hurricane season, the Citicorp building repairs were complete.

When it was over, the Citicorp building had become safer than most high rises, being able to withstand a 700-year storm. All the parties involved in this project were winners, including LeMessurier who, instead of suffering disgrace and financial ruin, ended his career on a spectacular success. He was respected and admired not only for his technical skill but also for his ethical behavior.

But what would have happened if the situation had been not so clear-cut? Suppose instead of a 16-year storm, a 50-year storm would have toppled the building? By the time this storm would most likely have occurred, all the principals involved with the building would probably have been dead. In that case, did they still have responsibility to the public? If so, how about a 100-year storm? At what point do we accept the risk and say that we simply cannot design the building to be any stronger?

(continued)

Box 4-12 (continued)

Kenneth Kipnis, in a famous article entitled "Engineers Who Kill" (1982), discusses the need for risk analysis in engineering design, pointing out (correctly) that engineering works often result in the death of people who use them. Kipnis believes that engineers cannot produce anything that is risk-free, but that engineers are beholden to the public to not subject them to risk levels above that which is commonly expected. For example, suppose the risk of having a backyard propane gas grill explode (using actual data) is one in 100,000 uses. The public is aware of this risk and by purchasing the grills, accepts this risk. To manufacture a grill that would explode once every 1000 times (without telling the public of the increased risk) is then clearly unethical.

Sometimes it is possible by simple engineering expediency to decrease risk. An example is the change in the hose connections for jet fuel and propeller engine fuel. Following an accident where a jet plane crashed because the ground crews had put the wrong fuel in the plane, the hose connections were changed so that it was not physically possible to hook the wrong fuel hose on to the airplane. Reducing risk by good engineering is one of the most satisfying aspects of the profession.

Box 4-13

Decision Making: Technical and Ethical Aspects

We need to distinguish between technical and ethical aspects of decision making. Technical questions are questions about how to achieve a goal by the most effective and efficient means, and there is usually (though not always) a state-of-the-art answer on which the acknowledged experts in the field agree—a right answer.

Suppose, for instance, that you own a computer company that has been asked to design a system for maintaining a centralized database that will enable the owner to quickly cross-match details of people who are entered in a database. You do not know the identity of your client—the arrangements are being handled by a third party, an international management consulting firm. The project presents interesting technical challenges because the client wants to have over 200 physically scattered sites networked to each other as well as linked to the central mainframe. The client also requires a complicated system of access at different levels by different users, which can rapidly be changed, and numerous safeguards against unauthorized access.

There will be right answers to all the problems that will crop up in the course of the project, and you and your employees will be able to come up with them.

Suppose that you somehow discover that the client is in fact a dictatorial and corrupt military government, which wants to install the system in order to keep track of political dissidents, especially supporters of a constitutional democracy. Such persons, even those who are merely suspected of supporting democratic ideals, are invariably arrested, tortured, and jailed for long periods in appalling conditions; many of them simply disappear without a trace. Let's say you believe that this is an evil regime. You now have to face a different set of problems—not technical, but ethical ones. Is it right for you to assist this government to repress opposition and mistreat its own citizens just because of their political views? Does it matter whether you work on it or not—since someone else surely will? What about your obligations to your employees, including the ones you hired specifically for the job? How important are your principles: Are you prepared to run the risk that your company will get a reputation for unreliability, and lose business?

These are not questions to which there is a state-of-the-art answer, on which experts agree. It isn't even clear that there *are* experts on ethics. I (Alastair) am an ethicist and I claim expertise in ethics, but I don't claim to be able to come up with the "right answers" to ethical problems. When organizations call me in as a consultant, I make it quite clear that my role is to help them to identify ethical issues and to resolve them in an honest and rational way that is consistent with the values of their society and that meets their own personal values, if these are more "demanding" than those of their society.

This doesn't mean that just any "solution" to an ethical problem is OK, so long as the decision maker is comfortable with it. It does mean that my clients have to face up to the hardest ethical issues, not ignore any of them, and be prepared to make some hard choices. There may be a number of resolutions to an ethical problem—unlike a technical problem, where there is usually only one best way—and some will be better than others. Even if ethics cannot come up with the *right* answers, it can and should be expected to provide *better* answers.

My clients (this is still Alastair speaking) often ask me for a litmus test of ethical decision making. I offer four possibilities:

- Can I look myself in the mirror for a full minute in the evening after I've made that decision, and feel proud (or at least not ashamed) of my decision? (This won't work unless you're honest with yourself.)
- Will I be comfortable if a report of my actions appears on the front page of tomorrow's newspaper? (This will work if the newspaper reports events accurately and honestly.)
- Can I tell my teenaged children exactly what I've decided, and why, and will they freely agree with my decision, or at least respect it? (This is a good test because teenagers are pitilessly moral and see through self-serving adult compromise. It won't work if you're an authoritarian parent and tell your children what to think.)
- Am I prepared to be judged by a panel of teenagers who are strangers to me—or by my students?

The building won't fall over, but you see the point. You ask Sarah what she thinks.

Looking you straight in the eye, she says, "Chris, you're a terrific engineer, and of course I'd back you all the way in any other circumstances. But I'm with Joe on this one. You'll just have to do the best you can in the time available. Hey, you've got nearly six months! If everyone works harder and smarter I'm sure you can fix it up."

Joe's administrative assistant appears, to remind him that he has to leave for another appointment. As you and Sarah depart, he says, "Aim for April 16 if at all possible. Worst case, we can get away with the two weeks delay they've allowed for, we can blame it on the weather, the city bureaucracy, or a subcontractor. But I want that hotel finished by April 30, not a day later. Just do it! Don't blow it! You're a great engineer, I'm counting on you! Make it happen!"

Rarely have you heard such a concatenation of managerial clichés: you're tempted to contact Scott Adams (creator of *Dilbert*) immediately.

Instead, you call Kelly and arrange to meet for lunch tomorrow.

Later the same day

That evening, you decide to call an old friend, Lee, whom you've known since high school. Lee is also a structural engineer, and you hope she'll be able to offer some advice.

"Hey, Chris, it's been a while. What's up? How's Alex and the kids?"

"Just fine. How's your brood?" Lee has four children by three different marriages.

"Noisy as ever." In the background you can hear what sounds like *The Simpsons*, Crash Bandicoot, and an electric guitar being tuned up, all at the same time.

"How's Sam?"

"Threw the jerk out. Went back to Reno, I gather."

"Surviving OK by yourself?"

"Yeah, better off without that loser."

You and Lee have an unspoken agreement that you don't discuss her taste in spouses.

"Can I talk to you about a situation in my firm, old friend stuff, OK?"

"Sure, what's the problem?"

"OK, we did this design for a hotel in Philadelphia, the Asmara. Neat structure, and it had a lot of interesting design problems. The way the architect drew the cantilevered structure over the front door is really interesting, it had to be analyzed using finite element analysis. My colleague Ali El-Sayd designed it, did you know him? He was killed in that terrible bus crash?"

"Ali? Yes, I met him at an ASCE local chapter meeting, he had just passed his PE[4] exam."

"That's the guy. Anyway, he apparently gave the overhang design job to Jerry, this is a young guy who had never worked with concrete before. He knew finite element analysis but he'd never been involved with a design like this one. After the accident, I checked out the design. It was wrong! Jerry used too few elements and most likely the re-bars in the overhang are improperly placed."

"But it's already been constructed, hasn't it? I saw it last week when I was in Philly, essentially completed."

"Yeah, that's the problem. It looks OK, but I don't know what happens if a strong wind or deep snow . . ."

"Whoa! Say no more! I don't want to hear it!"

---

[4]PE stands for Professional Engineer. Some engineers sign their name with the abbreviation "P.E." to signify their professional status, much like physicians sign with the abbreviation "M.D." after their name.

"Hey, what's the problem?"

Lee pauses, then says formally, "If this has anything to do with an ethical problem, and *especially* if it's a situation that might involve possible harm to the public, if you tell me, I have to report it to the State Professional Engineering Board."

"Oh, come on! We're just two friends talking."

"Chris, we've been friends for years, but when you tell me something that involves the health, safety, and welfare of the public, our friendship has to go in second place. I have to do this, don't you see? Don't tell me any more. See a lawyer, or go directly to the Board. Your very career might be at stake."

Box 4-14

Consulting with Colleagues

Lee's refusal to hear the rest of the story is not bad manners, nor is it an unwillingness to help a friend. Knowledge of a situation that can harm the public must be communicated to the proper authorities, and Lee would have had to rat on Chris. Even though professional colleagues are a great resource for professional advice of all kinds, the requirement to report unsafe situations overrides any friendship or collegiality.

Professionals usually support each other, and their primary professional identification is as a doctor, architect, or nurse rather than as an employee of an organization. Engineers, as noted earlier, are particularly collegial: Throughout their education, they're encouraged to work together and to help each other out, and this makes it doubly difficult for engineers to be critical of each other (Vesilind and Gunn 1998, 27).

Discussion Questions

4-1. Accepting that Dr. Kevorkian did not decide who should live and who should die, but just provided a service to people who requested it, was he a good doctor? Why or why not?

4-2. As professional service providers, is it ethically acceptable for lawyers to do whatever is necessary (while staying within the law) to help their clients to win, even if they believe that the result will be an injustice? Should they care about the ethics of their client's cause?

4-3. Do engineers face similar questions (question 4-2), and, if so, how should they resolve them? Is engineering different from other professions? If so, how?

4-4. While there is no question of the moral worth of engineering atrocities (see Box 4-10), other engineering projects are more morally cloudy. For example, was the atomic bomb an engineering success or failure? Can the engineers who worked on the bomb say that they were only doing engineering in the technical sense? The engineers would argue that they did not actually drop the bomb and therefore cannot be guilty of a moral crime. The mere construction of a device is not immoral (they would argue), but rather its use. Could the Nazi engineers use the same argument?

4-5. Does the moral obligation of engineers to society extend to the actual use of the engineered facilities or devices, whether this use is for moral or immoral purposes? Is the engineer expected to anticipate and consider the end use of the products of engineering skill? What do you think?

4-6. Consider the role of engineers in the construction of the low-cost housing in East Saint Louis, the Pruitt Heights buildings. Should the engineers have anticipated the impending social disaster? Engineers are not social scientists, so should they recognize impending problems such as incompatible lifestyles? Would you argue that the Pruitt Heights buildings were (a) an engineering failure, (b) an architectural failure, or (c) a planning failure? How much of the failure should be attributed to the tenants themselves?

4-7. Do you agree that ethicists cannot be expected to produce "the right answers" in the way that engineers are expected to do? Why or why not? (You may want to consult Gunn and Vesilind 1990.)

4-8. One of the reviewers of this book in the manuscript stage, an engineering ethics instructor, suggested that one solution for Joe and Sarah would be to write a letter to the owners, Timmo, stating clearly and forthrightly their concerns with the design of the overhang, and stating that in their opinion the overhang was unsafe. The reviewer suggested that if this letter was written, Pines (and Joe, Sarah, and Chris) were "off the hook," that there would be a paper trail of honest communication, and the decision to fix or not to fix the overhang would rest with the owner. If the owners wanted to ignore the problem and go ahead with the opening of the hotel, they were free to do so, but then the engineers would be blameless if the structure should fail. What do you think of this response? Do you think such a letter absolves the engineers of responsibility?

References

Freedman, M. H. 1966. "Professional Responsibility of the Criminal Defense Lawyer: The Three Hardest Questions." *Michigan Law Review* 64: 1469–1482.

Gunn, A. S., and P. A. Vesilind, 1990. "Why Can't You Ethicists Tell Me the Right Answers?" *Journal of Professional Issues in Engineering*, ASCE, 116 (1).

Kipnis, K. 1982. "Engineers Who Kill: Professional Ethics and the Paramountcy of Public Safety" *Business and Professional Ethics Journal* 1 (2): 75–90.

Mann, K. 1985. *Defending White Collar Crime*. New Haven, CT: Yale University Press.

McCollough, D. 1972. *The Great Bridge*. New York: Simon & Shuster.

Petroski, H. 2000. *Prism*. American Society of Engineering Education, November.

Report of the Presidential Commission on the Space Shuttle Challenger Accident (1986). Washington DC: June 6.

Searbruck, W. H., and F. J. Milliken. 1988. "Challenger: Fine-Tuning the Odds Until Something Breaks." *Journal of Management Studies* 25: 319–340.

Vaughan, D. 1996. *The Challenger Launch Decision: Risky Technology, Culture, and Deviance at NASA*. Chicago: University of Chicago Press, Chicago.

Vesilind, P. A. and A. S. Gunn. 1998. *Engineering, Ethics, and the Environment*. New York: Cambridge University Press.

Werhane, P. 1991. "Engineers and Management: The Challenge of the Challenger Incident." *Journal of Business Ethics* 10: 605–616.

5

Safety of the public

Saturday, October 19

"No way can you do that," Kelly exclaims. "Think how you felt after that bus crash—how can you even *dream* of doing that with innocent people's lives?"

"You did," you say defensively, having hoped to get some reassurance from Kelly.

"Hey, I didn't. When I argued with Arthur, he said it was just a bunch of fish and plants and stuff like that. He goes, 'It's not like we were going to kill people, and it's only for a while, until the new plant comes on line.' I don't go along with just killing off fish and plants as if they didn't matter, but I see where he's coming from.

Box 5-1

The Moral Status of Animals

In 1975, Australian philosopher Peter Singer (now a professor at Princeton) published his influential book *Animal Liberation*, in which he argued that our treatment of animals (more correctly, "non-human animals," since we are, after all, ourselves animals) is unjust, because we routinely use them as mere resources without regard to their interests. Singer builds his case for animal liberation on preference utilitarian grounds (see Box 2-9).

Conventional utilitarians take account only of *human* happiness or preferences, but, Singer argues, this is unjust because it arbitrarily excludes the interests of animals from consideration. He uses the term *speciesism* to characterize our refusal to count the interests of animals, on an analogy with racism and sexism, which systematically ignore the interests of ethnic minorities and women for the benefit of the dominant, white male group.

Everyone, including nonutilitarians, accepts that it is wrong to cause humans to suffer unnecessarily, because we believe that pain and suffering are an evil.

But animals are also sentient—that is, they can feel pain. Yet we cheerfully (or heedlessly) eat the flesh of animals, even though they may have suffered considerable discomfort, and sometimes considerable pain, in the process of being raised for food, transported, and slaughtered. Extreme examples cited by Singer include white veal calves (which are kept in tiny individual indoor pens in perpetual twilight and denied solid food so as to produce the white flesh favored by gourmets), pork and bacon (sows are kept perpetually pregnant and tightly restrained), and battery eggs (hens are crowded into small cages with uncomfortable sloping wire floors, painfully debeaked, and denied basic needs such as fresh air, dust baths, and the opportunity to form normal social groups). In these and other cases, the conditions of confinement preclude the animal's normal instinctive behavior, as well as causing physical suffering, in order to provide humans with plentiful and inexpensive supplies of meat and eggs. Certainly, many of us enjoy eating

(continued)

Box 5-1 *(continued)*

these products, but the price paid by the animals in pain and suffering is very high. Moreover, we do not need to eat animals or their products, as the large number of vegetarians amply demonstrates, and while meat is tasty and nutritious, there are equally tasty and nutritious alternatives that do not involve animal suffering. Singer even includes vegetarian recipes, which is certainly unusual (and practical) in a philosophy book.

Singer notes that we use animals for many other purposes such as consumer product testing, behavioral research, sport, and entertainment, where their interests are systematically violated for no significant human benefit. Even medical research involving animals, he claims, is rarely justifiable on utilitarian grounds. Much of it is a waste of resources, in that it does not result in any health benefits. He also points out that many of the medical conditions for which painful animal research is carried out are largely a result of human choices to expose themselves or others to unnecessary risks. For example, humans choose to smoke tobacco and thereby to run the risk of developing lung cancer: Why should animals be forced to develop lung cancer so that we can find cures for our own self-inflicted diseases when we could easily avoid developing them by not smoking? The same may be said of alcohol-induced liver disease, adult diabetes due to obesity, diseases caused by exposure to toxic substances in the workplace, and injuries suffered in road accidents, almost all of which are brought about by alcohol, excessive speed, carelessness, and in general the love affair that people in developed countries have with their cars. Had he been writing *Animal Liberation* today, Singer would no doubt have included AIDS, to the extent that it is self-induced by the individual's choice of a destructive lifestyle.

But what if the animal dies instantly, without suffering? Are you still harming it? Retired North Carolina State University philosophy professor Tom Regan has addressed this question in much depth, most notably in his 1983 book *The Case for Animal Rights*. Regan argues that many animals, specifically, "normal" adult mammals, have a valid interest in preserving their lives. Such animals are, in Regan's terms, "subjects of a life": Each animal has a life of its own, characterized by sentience, awareness of itself and the external world, a social structure, and the capacity for enjoyment. Thus, the lives of animals are valuable not just "instrumentally," as resources for humans, but for the animals themselves. As Regan puts it, they have "inherent value," just as humans do, and it is therefore appropriate to ascribe rights to them, just as we ascribe rights to humans.

Of course, most humans are more intelligent and rational than any animal, but even humans who fall well short of our standards of "normality," such as severely brain-damaged infants, are still accorded rights. We would not dream of doing cancer research on "human vegetables" or even on people who are in what is called a "persistent vegetative state" (see Box 4-2), yet we willingly inflict painful, often fatal diseases on alert, intelligent chimpanzees. This argument, often called "the argument from marginal cases," is intended to show that we are inconsistent in failing to respect the rights of animals that have to a high degree the qualities that we consider make humans so special, even while respecting the rights of humans who do not have these qualities.

According to Regan, therefore, even if you could guarantee to kill an animal instantly and painlessly, you would still be failing to respect its right to life, which is based on the fact that as a being with a life of its own, it has inherent value.

Perhaps conventional ethics undervalues animals by accepting our treatment of them as mere resources. Ethicists refer to the idea that only human interests matter as *anthropocentrism* (the root is the Greek word *anthropos*, meaning a man or human being in general). For *anthropocentrists*, the moral community (those beings whose interests count and whose treatment is a matter of ethical concern and obligation) includes only humans.

Not all ethicists are anthropocentrists: Singer and Regan are not, because they believe that we ought to respect the interests of animals and value them as more than mere resources.

Environmental philosopher Paul Taylor, formerly professor of philosophy at Brooklyn College, has argued for "biocentrism," the view that all living things ought to have a place in the moral community or are "morally considerable" (1986). In some ways, Taylor's view resembles that of Regan who, you will recall, states that many animals have "inherent value"

(continued)

Box 5-1 (continued)

and are entitled to respectful treatment because they have a good of their own. Taylor (who uses the term *inherent worth*) would have us extend respect to all living things, including plants because they, too, have a good of their own. A tree, for example, will thrive or sicken and die, depending on its environment, which includes our treatment of it. A machine may also be said to be well or badly treated: For example, the condition of your car depends on how well you take care of it. But unlike a living thing, a car does not in any sense have a good of its own. It may function more or less well, but it cannot flourish or fail to flourish. Ethically, the difference between a tree and a car is that the tree is a living organism, which has a value that is independent of its value as a resource, whereas the car is only a thing, a means to an end, and has only *instrumental* value. In Taylor's view, this difference qualifies the tree, but not the car, as morally considerable, as worthy of moral consideration.

Arthur (Kelly's boss) adopts an anthropocentrist perspective, as is clear from his dismissive reference to "just a bunch of fish and plants." Arthur would presumably distinguish his firm's actions from those of Pines, which put the lives of people at risk. It seems that the only reason Arthur would condemn water pollution would be if it appeared likely to harm humans, for instance by contaminating water supplies or by threatening the livelihood of fishers. William Baxter, in his well-known book *People or Penguins: The Case for Optimal Pollution* (1974) develops a case for environmental policy to be based purely on considerations of the net benefits to humans. "Penguins, for example, are valuable only because people enjoy seeing them walk about on rocks" (p. 5). If they didn't enjoy watching penguins, there would be no reason to preserve them.

"Optimal pollution" sounds almost like a contradiction: If pollution is bad, how can it ever be "optimal"? However, Baxter believes that some pollution is an inevitable consequence of, for instance, the production of consumer products. We are prepared to accept some pollution in order to have the benefits of dishwashers and washing machines; we would not accept reduced pollution levels if as a result we did not have adequate access to these goods. Thus, we balance levels of material comfort and environmental quality. The point at which we decide that the production of more appliances will raise pollution to an unacceptable level is the point of "optimal pollution."

"Anyway, it's all worked out because the day after I told my boss Arthur I'd ignored the chromium problem, the governor announced a major crackdown on river pollution, and next day the state EPA said it was going to tighten up the regulations for discharge permits including both kinds of chromium, so we wouldn't have been able to get away with it anyway. Like, Arthur is *dying*. But I get points for loyalty *and* good engineering work. And I get to do the right thing and keep my job."

Box 5-2

Ethical and Legal Obligations

Kelly got out of her ethical problem because the state passed new environmental laws, so she was able to insist that the discharge be treated. This suggests that Arthur thinks that being ethical is no more than conforming to legal requirements. This is quite a common position in business ethics. University of Chicago economist Milton Friedman is well known for his advocacy of the view that managers' sole obligations are to the stockholders of their company, to maximize their return on investment by maximizing profit. The only constraint is the requirement to keep within the law and normal business practice. In Kelly's case, there is an obligation to meet discharge standards, but there is no obligation to go further. In fact, according to Friedman, if you voluntarily improve your company's environmental performance beyond what is legally required, you act wrongly because the extra expenditure is effectively a theft from the owners.

"Lucky you," you say bitterly. "There's no way I'm going to get out of this that easily. It's a real dilemma! I can do some remedial work on the hotel, but there's still going to be a risk of a major disaster. I mean, it quite possibly won't happen, but what if it did? And I'd be worrying all the time that it might."

Box 5-3

Ethical Dilemmas I

In everyday language, a dilemma is a problem. But as ethicists use the term, a dilemma is a particularly difficult situation in which neither of the two alternatives available is ethically acceptable, usually because each would require the sacrifice of an important principle. For instance, suppose you're a psychologist faced with this situation: You're treating a client who admits he is physically abusing his kids. He wants to stop—that's why he consulted you—and you expect to be able to help him. Meanwhile, you're concerned for the safety of the children, but you're torn between two duties. You assured your client that his confidentially would be respected, before you agreed to help him—that's standard practice—so you owe him a duty. But you also feel a responsibility to protect the children from further harm even though they're not your clients. In other words, you want both to protect the children and to honor your obligations to your client, *but you can't do both*. If you say nothing, maybe one or more of the children will be seriously hurt or even killed. You can prevent that by informing the child protection authorities, but they'll probably take the children into care, and the client will probably go to jail. Prisons rarely have adequate counseling help available to inmates, and even if they do, your client probably won't trust professionals again. Some philosophers call such a dilemma a *tragic moral choice*. No matter what you do, you lose.

Fortunately, most of us will never be confronted with such tragic moral choices. If we are, we can get help from counselors who understand the dilemma. A job of ethicists is to help such people find acceptable alternatives to the unacceptable options that they have considered. In the case just discussed, a better alternative would be for the abusive person to move out until the therapy has successfully changed his or her behavior.

We have more to say about these issues later (Box 9-4).

"So, tell Joe and Sarah you won't try the Band-Aid bit."

"They'll fire me."

"I don't think so. Worst case, they'll give you a while to find another job—no problem. Sure, you'll probably lose some seniority, but in a year or two you'll be back where you are today. You a professional or what? Maybe you'll be behind 10, 20 K, but truly, how much are those lives worth?"

Box 5-4

Calculating the Value of Life

Calculations about the value of human life are commonly made by agencies such as the U.S. Department of Transportation, as part of a cost-benefit analysis, in order to decide whether to spend money on highway improvements. If a certain improvement costs, say, $1,000,000 and is expected to save two lives a year, it is worth doing it only if the value of a human life is worth more than $500,000. How is this calculation done?

To oversimplify, suppose that the average age of fatally injured persons is 40 years. This person would have 25 years of employment remaining at an average salary of, say, $30,000 per year. Thus, he or she is worth $750,000,

(continued)

Box 5-4 (continued)

so the improvement should proceed, unless a competing project has a more favorable cost-benefit ratio.

Many people would consider this to be cold and unfeeling, treating a person as no more than a salary-earning machine. Notice that we would not take this view if the other items such as physical damage to motor vehicles, traffic delays caused by all accidents, costs of emergency services, and so on, are figured in. It makes sense to spend $50 on a warning device on a piece of machinery that would remove a 10% risk of $1000 of damage being caused by a malfunction, and that is because the cost-benefit analysis is favorable. It makes sense to install a security system in your home if the saving on your insurance premium is significantly greater than the cost of the system. But can we put a value on human life?

Many people would deny this. A recent study in New Zealand—which has one of the highest rates of youth suicide in the world—argued forcefully that more resources should be spent on suicide prevention in young people, including improved mental health services. No one disputes this—the suicide of a young person is a terrible tragedy. But many people were shocked that the study also undertook to show that youth suicides have an economic cost too. The death of an 18-year-old, in whom society has made a considerable investment through education and health care, represents the loss of many years of income. Moreover, only one in ten suicide attempts is successful, leading to considerable health costs that in the New Zealand health care system are mostly funded by the taxpayer. People asked: What has this got to do with it? We're talking human tragedy here, not loss of earnings, wasted investment, and health costs.

Defenders of cost-benefit analysis involving human life argue that unless we put a monetary value on life, then we cannot decide what its value is, because the dollar is the only neutral and quantitative method of valuing anything. Since we all agree that human life is indeed valuable, the alternative to assigning monetary value is to say that each life is priceless—or worthless! But then we would have no basis for deciding how to allocate limited resources. The priceless, being beyond price, cannot be compared with other valuable things that do have a price, and thus they cannot figure into the equation. Transportation authorities, perennially strapped for cash, have to allocate resources where they will do the most good.

Defenders of cost-benefit analysis also argue that as a matter of fact we do accept decision-making tools that put a limited value on human life. Most people do not believe that every seriously ill person should be kept alive for as long as possible regardless of their quality of life (see Box 4-2). But once we say that we are deciding that some people's lives are not worth maintaining—not only because the life is burdensome to them and those around them but because it is a futile use of scarce resources—most of us would vote for health resources to be spent on suicide prevention rather than on keeping irreversibly comatose patients alive indefinitely. Even if we don't put an actual dollar value on life, we are still saying that more should be spent on keeping some people alive than on others. Thus, we are making a kind of cost-benefit analysis.

"OK, but Pines will just find someone else to do it."

"Sounds like Pines Engineering needs some help in becoming ethically responsible."

"Yeah, I suppose so, but it's not my responsibility to make the firm more ethical."

Box 5-5

Fix Up Your Organization Ethically?

James Muyskens, in his book *Moral Problems in Nursing: A Philosophical Perspective* (1982), argues that the roles of a nurse as employee and as professional are sometime in conflict. As an employee, "the nurse is hired to carry out the directives of the physician and to support the policy of the hospital administration . . . yet, on the other hand, the nurse is legally and morally accountable for her or his judgments exercised and actions taken" (p. 158). Individually, a nurse is powerless, while collectively, the profession has great power. "If all nurses were to walk out tomorrow, the system

(continued)

Box 5-5 (continued)

would collapse" (p. 159), and, Muyskens believes, the profession as a whole has a collective responsibility for improving patient care in inadequate health provider organizations. However, because the individual nurse is relatively powerless, she or he cannot be blamed for the employer's low standards.

Unlike nurses, engineers—even quite recently qualified engineers—are often in a position of responsibility for projects. Taft Broome, a professor of civil engineering at Howard University (and an ethicist), tells the story of his first day on the job at a construction site (1999). The site engineer and the foreman both disappeared and left Broome alone. A convoy of concrete trucks pulled up, and the driver asked Broome where to place the concrete. Having no idea what the concrete was for, Broome told the driver to take it back. The driver explained, no doubt with great disdain, to the young engineer that the concrete would have to be dumped somewhere, otherwise it would harden in the truck. If Broome did not tell him where to pour it, he and the other five trucks would just pour the concrete on the ground. Faced with a critical decision, Broome went into the construction trailer and in a few minutes figured out from the construction log and plans where the concrete was to go. Although his problem was not a life-or-death situation, it demonstrates how even young and inexperienced engineers are asked to assume responsible positions and make critical decisions. Broome could not pass the buck in his role as the (junior) engineer.

Another example of a young engineer being asked to make a major decision occurred in a case that is still in court at this writing (and therefore cannot be further identified). A major hazardous waste landfill was under construction, and the contractor was constructing a seepage barrier that was to prevent groundwater from flowing into the landfill. In order to save money (and hence enhance his own profit), the contractor asked the junior engineer on the construction site if it was OK to place some rubble into the fill area. The rubble would eventually have had to be buried anyway, argued the contractor, so why not use the rubble instead of importing special soil from off-site.

The junior engineer passed the request up to his boss, who agreed that this might not be a bad idea. The boss also suggested that using the rubble seemed to be a reasonable thing to do. Besides, "Once it is buried nobody will know what is under the ground." In addition, the senior engineer told the young engineer that this contractor has a tradition of sharing his profits with the engineering firm in the manner of Christmas gifts of a substantial nature. The senior engineer made it very clear that these "gifts" were much appreciated by the engineering firm. But he told the junior engineer that the final decision was his.

Wanting to be a team player and not wanting to make waves, the junior engineer decided to allow the contractor to use the rubble.

Soon after the landfill was finished, the monitoring wells began to show significant groundwater seepage and the story came out about the placement of the rubble. Large chunks of concrete cannot, of course, be well compacted and will leave huge channels through which water can flow. The landfill barrier had to be dug up, at great additional expense, and properly constructed.

As would have been predicted, everyone blamed everyone else for the leaking landfill. Most notably, the junior engineer was blamed for signing off on this change in design, even though his bosses had in effect agreed to the substitution. The senior engineer argued that the final decision had to be made in the field, and that this was the responsibility of the junior engineer. They never mentioned anything about the "gifts."

In this case, the junior engineer was not powerless to stop what he suspected was a violation of engineering ethical principles (not to say good hydraulics). If he had been given a direct order from his boss to allow the substitution, then he would have had a difficult decision to make, but in this case the decision was technically wrong (both engineers should have realized that the rubble fill would conduct water) and ethically cloudy. The suggestion by the senior engineer that the contractor is generous with his Christmas gifts should have tipped off the junior engineer, and all kinds of red flags should have gone up. He should then have explained the situation to the contractor, and more than likely that would have been the end of the matter. Unfortunately, he made a bad decision within a corrupt organization.

"I hear what you're saying. Something bad happens, at least *you* won't be responsible. You've tried to do all you can."

"Yes, I will be responsible, because I'll *know*. The only way I can get rid of that moral obligation is to go directly to Timmo—or the City of Philadelphia Building Inspectors."

Kelly thinks about this.

"You're right," she admits. "But don't rush at it. Whistle-blowing's *serious*. I'll fax you this cool article I just read, guy called DeGeorge wrote it."

Box 5-6

Whistle-Blowing I

Gene James (1988) defines whistle-blowing as "The attempt by an employee or former employee of an organization to disclose what he or she believes to be wrongdoing in or by the organizaton . . . an effort to make others aware of practices one considers illegal, unjust, or harmful."

There are many forms of whistle-blowing, ranging widely in their effect and risk to the whistle-blower. The least-dangerous kind of whistle-blowing is going over the head of one's immediate superior. But even this is fraught with danger. If the superior is then criticized by his or her bosses, then you will be blamed for it and there will be some unhappiness. If going over the head of the immediate superior does not resolve the problem, the whistle-blower can keep going up the chain of responsibility within an organization, incurring increasing displeasure within the organization. This type of whistle-blowing is called *internal* in that the information does not leave the organization.

If no satisfactory result is obtained within the organization by internal whistle-blowing, then going outside is an option, in which case the whistle-blower becomes an *external* whistle-blower. This will more often than not result in getting fired, with significant disruption and personal loss. Whistle-blowers can also be either *anonymous* or *named*. Writing an anonymous letter to a newspaper often (but not always) protects the identity of the whistle-blower, but such accusations often are not taken seriously. The strongest letters are ones that are signed. Finally, whistle-blowers can be *active* or *alumni*. One can continue to be employed by the organization (and then be fired), or one can quit the organization and then, as an alumnus, blow the whistle.

The decision to blow the whistle, from the mildest to the most severe form, is a wrenching and difficult personal decision. Richard DeGeorge (1981) tries to answer two questions: When is whistle-blowing *justifiable* (it is ethically OK to blow the whistle) and when is it *obligatory* (one has a duty to blow the whistle)? DeGeorge states that he is writing only about the private sector, but we can see no reason why it should not apply to government and NGO[1] employees too.

On the first question, he argues that external whistle-blowing is *justifiable* if the following four conditions are met:

- The practice is causing or is about to cause *serious harm*.
- The employee has reported the concern to his or her *immediate manager*, without success.
- The employee has exhausted the procedures that are internal to the organization, including reporting the concern to *top management*, without success.
- The employee believes that the good that will be achieved will outweigh the harm that will be caused, both to the organization and to the whistle-blower himself or herself.

Most people would probably accept these conditions. Whistle-blowing has the potential to cause serious damage to an organization's reputation, as well as subjecting the employee to stress and hounding by the media, and putting his or her career at risk.

Secondly, DeGeorge argues that whistle-blowing is *obligatory* if these conditions are met:

(continued)

---

[1]Nongovernmental organizations, such as the Red Cross, the Boy Scouts, or the Sierra Club.

> ### Box 5-6 (continued)
>
> - The employee has adequate *documentation*, that is, documentation that would convince a reasonable (and, presumably, adequately informed) person.
> - The employee has good reason to believe that it will be *successful* in remedying the situation.
>
> Again, these conditions seem reasonable. If the employee doesn't have documentation as objective evidence of the alleged wrongdoing, she or he is unlikely to be able to make a successful case, and unless there is at least a reasonable chance of success, the employee surely cannot be expected, as a matter of ethical obligation, to risk his or her career.
>
> DeGeorge also states that, contrary to Kenneth Alpern's claims discussed in Box 2-2, we should not accept "the myth of the engineer as moral hero." He continues, "We cannot reasonably expect engineers to be willing to sacrifice their jobs each day for principle and to have a whistle ever at their side" (p. 1). This issue is discussed further in the next box.

"And I think you should talk to a lawyer. I know this one lawyer, Shawn Roe, who's done a lot of work on engineering liability. He really knows his stuff."

By the time you get home, Kelly has already faxed you the DeGeorge article. You read it carefully and discuss it with Alex, who suggests that according to DeGeorge's criteria, you'd be justified in blowing the whistle but you wouldn't have an *obligation* to do so. "It all depends on how much you stand to lose, and whether you're prepared to lose it."

"But I have to think of you and the kids, too."

"Yes," Alex says. "But my job can easily support us if you lose yours, and even whistle-blowers get jobs eventually. Anyhow, I'd never pressure you to do anything—it's your choice."

> ### Box 5-7
>
> ## Whistle-Blowing II
>
> Whistle-blowers commonly suffer for their actions. Stress, criticism from colleagues, isolation, and hounding by the media are probably inevitable. Some whistle-blowers have been fired—especially in the United States where employers have the right "to hire and fire at will," that is, without having to show cause. In highly unionized industries such as brewing and automobile manufacture, employees usually have contracts that preclude arbitrary dismissal, but this is rarely the case with engineers.
>
> Whistle-blowers have sometimes suffered very serious consequences—their careers and even their lives have been wrecked. The movie *The Insider* tells the story of Geoff Wigand, head of research at tobacco company Brown and Williamson, who suffered marital breakup.
>
> In *Silkwood*, it is even claimed that an employee of a utility company, Karen Silkwood, was alledgedly murdered by her employers while she was on her way to deliver a journalist papers that, allegedly contained details of violations of safety standards in her employer's nuclear generation facility. Myron Glazer reported the fate of a number of whistle-blowers in an article, "Ten Whistle-blowers and How They Fared" (1983).
>
> Patricia Werhane (1988), while describing such acts as supererogatory, concludes: "The bottom line is that each of us is responsible, morally responsible, when we have requisite knowledge and the possibility to choose. Taking moral risks is part of exercising that responsibility."
>
> *(continued)*

Box 5-7 (continued)

Some companies do accept that employees may sometimes be justified in blowing the whistle. For example, the engineering firm of Frees and Nichols, in the introduction to its Ethical Conduct Policy, states that "The foundation of a professional firm such as Frees and Nichols is ethical practice," and the Policy itself includes the following statement:

> Disclosure—Whistle Blowing—If an employee feels that an unethical or dangerous condition exists, either on a client's project or anywhere else in our operations, the employee shall (note: not may) report this to their supervisor immediately. If the employee is uncomfortable with this particular person, the next higher level or another higher manager may be contacted. Under no circumstances will an employee be disciplined for calling attention to perceived problems. Such concern may be submitted, anonymously, if there is a concern about retribution. However, such actions should be thoroughly pursued within the company before any public disclosure is considered. (Quoted in NIEE newsletter, May 1998)

Monday, October 20

You call Shawn Roe's office first thing and, as luck would have it, he has had a client cancel an appointment and is free to see you late in the afternoon.

The attorney turns out to be a well-built, middle-aged man with an uncanny resemblance to Hall of Famer Joe Morgan, the former Oakland third baseman turned ESPN commentator, right down to the sharp suit and explosion-in-a-paint-factory silk tie. After you've introduced yourselves, he says, "Before I begin, let me tell you two things. First of all, our conversation is protected by legal confidentiality, and nothing you say will be repeated outside this room without your permission."

"I understand that."

"Good. But secondly, a lawyer is not allowed to lie on behalf of his or her client, nor to encourage or assist a client to commit perjury, to lie to a court. Please bear this in mind when deciding what to tell me."

"OK."

"So, tell me your story."

You describe the situation in full, ending with the discussion with Lee.

"So is Lee right? If I had explained the problem, would it have been necessary for Lee to go to the Professional Engineering Board?"

"Yes," Shawn says. "Technically that is right. If something terrible were to happen and you were caught in the investigation and revealed that you had told Lee about the problem, both of you could lose your engineering license. In fact, the same regulation governing such behavior seems to govern yours. Under the State Administrative Code for the Registration of Professional Engineers, if you do not report the faulty design, you are guilty of professional misconduct. Let me read paragraph 701(g)2 to you.

"701(g)2 If he ('Isn't that amazing!' he inserts. 'They still have all this sexist language in the Code.') has knowledge or reason to believe that another person or firm may be in violation of any of these provisions ('That is, all the above guidelines to practice'), he shall present such information to the Board in writing and shall cooperate with the Board in furnishing such further information or assistance as may be required by the Board.

"Note the word 'shall' in that statement, Chris. It means that you *have* to do it or you can be reprimanded by the Board or even lose your professional engineering license."

"Sure, but if I do nothing and the overhang doesn't fall, then nobody knows any better, and I'm off the hook."

"Yes, then you're off the hook."

You think for a moment. "Sounds like there are three possibilities. I tell the Board and they make a stink and the overhang gets fixed; I don't tell the Board and it doesn't collapse; I don't tell the Board and it collapses, maybe killing people."

"That pretty much summarizes it."

There is a lengthy silence, undoubtedly an expensive silence.

"You had cases like this before, Shawn?"

"Yes, several."

"Did any of your clients opt for not telling, with disastrous results?"

"None in which people were killed, thankfully, but yes, I had a case where my client, in spite of my urging, decided not to tell the State about a hazardous waste spill. She insisted it would soon dissipate and then nobody would ever know. Unfortunately for her, and purely coincidentally, the town did some groundwater monitoring in anticipation of a landfill being located near the property and discovered the contaminated groundwater."

"What happened then?"

"The State was notified and went directly to the oil company she worked for, who told them that all of their technical work was being handled by their consulting engineer. It was obvious to everyone that the engineer knew about the spill and failed to report it. The company shrugged its collective shoulders and said that they must have been misled by their engineer. This was an unlawful act, under the federal Resource Conservation and Recovery Act, and the engineer was criminally liable. She was convicted and paid a handsome fine. She avoided jail time by a plea bargain for community service."

"She kept her engineering license?"

Shawn gives you an old-fashioned look. "Are you serious? As soon as the criminal proceedings were over she got a fat envelope from the State Board of Professional Engineers requesting all the documentation on the case. They were about to open their own hearings. The Board waits for legal proceedings to develop and then they step in, using all the available documentation from the previous litigation. Anyway, the Board charged her with gross violation of the Standards of Professional Conduct. Specifically, the very first statement of the Code says: 'The engineer shall hold paramount the health, safety and welfare of the public.' She didn't have a chance. They nailed her. But it took a year of hearings and letters and witnesses, some of the same ones that gave testimony in the criminal case. It was a bloody mess. And expensive. I don't want to tell you what my bill was. Bottom line, she lost her PE license. Revoked, no opportunity to appeal."

"Talk to me as a counselor. What should I do?"

"Chris, none of your options are easy ones. I can tell you though that if that structure collapses and people are hurt or even killed, you will be hung out like a sacrificial lamb. Everyone will point to you as the engineer who knew the structure was unsafe and who refused to do anything to stop a possible disaster from occurring."

"But . . ."

"Have you exhausted all options within the firm?"

"Yes. I went to the president. I even sent him a letter to make sure he understood the problem. Nothing has happened."

"Then you have three options (other than the infamous 'do nothing' option of course). You can first go to the city building inspection department. They'll want to know how you know all these things and so on and so on. You cannot remain anonymous. Most of the inspectors are not licensed engineers and you can overwhelm them with technical jargon. You can get their attention by telling them that they have a potential disaster on their hands. This they will listen to. It's their job to make buildings safe."

"Can they stop a job? I've never heard of that."

"Oh yes, they can, and they will. Simply revoking the construction permit, or not giving them an occupancy permit, they can make life very miserable for the owner and contractor."

"But my firm will know I talked to the building inspectors."

"Probably. The inspectors will need professional help in resolving this issue. Your concerns will be on the table."

"So I'm dead in the water. Goodbye career."

"Not necessarily. It could be that the issue can be kept out of the papers and resolved right at the city engineer's office. Nobody wants to crucify anyone. There has been no disaster. Nobody has gotten hurt. All anyone knows is that the job has been halted. It can be kept out of the media. Remember this is a technical disagreement, not a legal one."

"If the owner wants to make it a legal one?"

"Yeah, he can. Then it will become public. But more than likely, the owner will want to keep it out of the media also for fear of bad public relations. He absolutely does not want his hotel to be thought of as unsafe—certainly not another Hyatt Regency disaster. But if the opening is postponed and he loses a lot of business due to the delay, he will come after the engineering firm with a vengeance once the building is complete. Hope they have good insurance."

Box 5-8

Disaster in Kansas City

In 1981, the walkway in the atrium of the Kansas City Hyatt Regency Hotel collapsed, killing 114 people and injuring 186. This was the most devastating structural failure in terms of human life ever to occur in the United States.

The walkways, as shown below, were suspended from the ceiling and were used for pedestrian traffic.

On the 17th of July, the hotel was hosting a party and nearly 2000 people were in the lobby. A band was playing and people were dancing on the walkways. They noticed that the walkways appeared to be wobbling, but thought nothing more of it. The collapse came without warning when the fourth-floor walkway *(continued)*

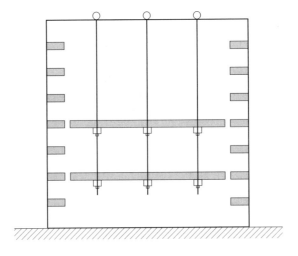

Box 5-8 (continued)

came crashing down onto the second floor walkway and both of them crashed to the lobby floor.

An architectural firm designed the hotel, and the owner hired the engineering firm of G.C.E. International to work with the architect in producing the design drawings. The president of the firm was Jack Gillum, P.E., who assigned the job to Daniel Duncan, P.E. The final drawings for the structure were prepared by Duncan and sealed by Gillum, as is the custom. The contractor for the steel fabrication was Havens Steel Company.

As the final construction drawings were being prepared, Havens engineers contacted Duncan to tell him that the suspension for the walkways was not buildable. The walkways were to be suspended by six rods hanging from the ceiling, and these rods were to carry both the second- and fourth-floor walkways, both suspended from the same six rods. The way the architect had sketched the drawings, however, made it impossible to construct. As shown by the detail of the original connection to the fourth-floor walkway, the continuous steel rods were to extend through the fourth-floor walkway, and the walkway was to be held up by six nuts threaded on the rods. The thread, however, was only where the nuts were and the rest of the rod was to be smooth. It would be impossible to get these nuts into place unless the entire rod was threaded or unless the rod was milled to smaller diameter so that the nuts could be slipped through the rod to the threaded sections. Because both of these solutions were either prohibitively expensive or structurally unsound, the Havens engineers suggested that they use two different sets of rods, one set extending from the ceiling to the fourth-floor walkway, and one set from the fourth-floor walkway to the second-floor walkway. Duncan agreed and revised the design, which was then approved by Gillum. The new connections on the fourth-floor walkway are illustrated in the drawings.

This design had two serious flaws, however. First, the use of two steel channels to hold the nut without having a bearing plate (large washer) appeared to be against the Kansas City building code. But more seriously, the nuts holding up the fourth-floor walkway now had to carry not only the weight of the fourth-floor walkway but also the weight of the second-story walkway, because the second-floor walkway was now to be suspended from the fourth-floor walkway. While the construction of the nuts without a bearing plate was already dangerous, the fact that the load was now doubled made the design radically unsafe.

(continued)

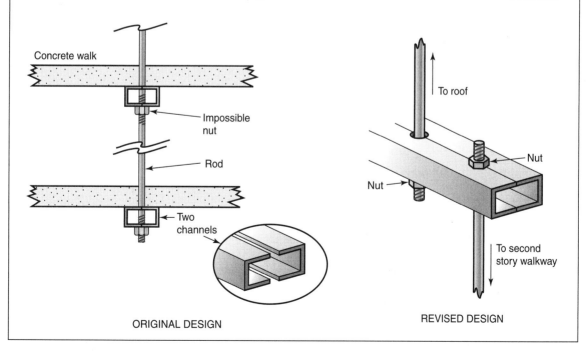

ORIGINAL DESIGN

REVISED DESIGN

> **Box 5-8** *(continued)*
>
>
>
> After the walkways had been constructed, the construction crew sensed that something was wrong, and many of them refused to walk on the second-floor walkway for fear of collapse, nor would they run their wheelbarrows over it. There is no evidence, however, that these concerns were expressed in any formal way.
>
> When the party was going on and people were dancing on the walkway, the nuts on the fourth-floor connections experienced not only the dead load of both walkways and the live load of the people but also the dynamic load of the dancers, and the connections failed.
>
> Who is to blame? Clearly, the architect did a stupid thing by designing a structure that could not be built. But architects rely on engineers to figure out how to build what they design, and to build structures that are safe. The initial engineering error seems to have been Donald Duncan's, who did not realize that he was doubling the load on the nuts by going to the two-rod system. There seems to be evidence that he did not take the diligent care with the design that he ought to have. The commission reviewing the appeals of the engineers noted that
>
>> the level of care required of a professional engineer is directly proportional to the potential for harm arising from his design and . . . indifference to harm and indifference to duty are closely related if not identical. (*Duncan v. Missouri Board*, 1988)
>
> Engineers are not expected to be perfect, of course, but only to exercise "reasonable care and competence" (*Gagne v. Betram*, 1934). The designer of the Tacoma Narrows suspension bridge, which collapsed in 1940 due to wind-induced oscillations, was not viewed as negligent. The bridge, built in 1938, was at the time the most slender suspension bridge ever built, but it was believed to be a reasonable construction by the generally accepted standards of the day. The level of care should have been great when designing the connections of walkways that could collapse and kill people, but apparently this was not done.
>
> The construction drawings, with the faulty connection, were sealed by Jack Gillum, who was ultimately responsible for the design and for the collapse. He looked over the drawings and also did not catch the mistake. In his case, he compounded this error by trying to argue that he was not, in fact, responsible for the collapse. He asserted that even though he had sealed the drawings, it was the fault of the architect, the steel fabricator, and Donald Duncan, but not him. The fact that he tried to blame others for his oversight was actually an admission of guilt. By sealing the drawings, he certified that they were correct. By then blaming others, he admitted that he made a mistake in sealing the drawings. Both ways he loses. Such behavior is patently unprofessional and unseemly. The commission came down hard on Gillum:
>
>> Gillum was by statute responsible for those drawings and he accepted such responsibility when he entered into the contract and utilized his seal. His refusal to accept a responsibility so clearly imposed by the statute manifests both the gross negligence and unprofessional conduct found by the Commission. (*Duncan v. Missouri Board*, 1988)
>
> Both Duncan and Gillum were talented structural engineers whose careers were ruined by this tragedy. But this is what engineers are paid to do—to take responsibility. This is why they have sleepless nights worrying about what adverse effect might result from their works. Nobody said that being an engineer was an easy job.

"What other options are there?"

"You can go to the State Board of Professional Engineering. They provide an avenue to air your concerns without going public. In fact, if you can't get your concerns resolved, you are required to go to them, as we discussed earlier. They have no power to shut down the job, but they will send out a lot of letters and make a lot of stink. Of course, you will be in the center of it. But the firm will deny everything and pretend that the problem never existed. I have even known firms to go back and change the 'as built'[2] drawings to avoid litigation. Dangerous stuff, that. Tampering with evidence."

"Omigod. What's my third choice?"

---

[2]"As built" drawings describe the project as it has been built. Every project is changed during its construction due to unforeseen circumstances, and at the conclusion of the project, the engineers prepare drawings of the project as it actually was built. These drawings are used by future engineers in designing modifications and additions.

"You can go directly to the newspapers. Do not pass "GO," do not collect $200."

"And then pack my bags for Tibet?"

"Something like that."

"I'll have to think about it. Maybe the problem isn't as bad as I thought. Maybe I made a mistake. Maybe the overhang will be just fine. Why should I be such a pessimist and expect the worst? After all, some of my best professional friends believe there isn't a problem. If I do something rash I'll lose my job, or worse. I'll have to think about it, yeah."

Silence.

"You know, they never told us any of this stuff in engineering school."

"Maybe they thought ethics was not a part of professional engineering."

"Yeah. How wrong they were."

"So what do you think you'll do?"

Box 5-9

Options

What courses of action are available to Chris? One would be to do nothing and just go along with the opinions of colleagues. After all, Chris may not be right after all. No doubt Chris is feeling less than omniscient, worried that maybe there is an error in the calculations. And maybe there is no real problem. Maybe the overhang was poured in such a way as to increase its factor of safety and enable it to withstand all the wind and snow loads.

Alternatively, if Chris believes the calculations, resignation might be an option. Or maybe working on a different project. While this would alleviate immediate responsibility, it still would not reduce the moral obligation because Chris *knows* the overhang is unsafe.

If all options within the organization have been exhausted or rejected, then going outside the firm is a choice (external whistle-blowing).

Chris could go to the local building inspector. All structures within a community in the United States (and many other countries) require both a building permit and an occupancy permit. The building permit is issued before construction can begin. Typically, the local city engineer will go over the plans with the design engineer and do a cursory analysis, not a full structural analysis. Then when the building is finished, before occupancy can occur, the city engineer will go around and inspect the facilities, looking at plumbing and lighting, mainly to make sure that everything works and there are no dangling wires or broken sewer lines to cause public health problems. The city engineer has no idea whether the structure is safe or not. Only the design engineer knows this.

Suppose Chris goes to the building inspector and expresses the opinion that the overhang is unsafe. The inspector will be worried and will call in the design engineer from Pines.

"I understand you did a quick and dirty finite element analysis on the overhang at the entrance," the inspector will say. The Pines design engineer will know immediately that Chris must have been talking to the inspector because it is highly unlikely that the city engineer will have any idea what a finite element analysis is, let alone that it was done incorrectly. So the cat will be out of the bag. The design engineer will have to show him or her the calculations, and the city engineer will probably call in a consultant (perhaps a university professor) to help evaluate the design. The flaw will be discovered, and in this scenario, completion of the building will be held up. There will be considerable extra construction costs that will have to be paid by Pines.

On the other hand, suppose Chris goes to the client with the concerns. One of two things will happen. The client may do nothing, in which case the only recourse is to go to the building inspector. Or the client will contact either Joe, Chris's boss, or the design engineer at Pines, which will point the finger even more conclusively at Chris since clients know absolutely nothing about structural engineering.

In short, there seem to be only two options: Do nothing, and ignore the Code of Ethics, or blow the whistle, thus meeting ethical obligations but running a considerable personal risk.

Are there other alternatives? What else might Chris consider doing, and would these options be any better?

Monday, October 21

You don't like any of your options, and you don't want to be a moral hero, so you've decided to duck the whole issue and ask to be assigned to another project. Given your unique circumstances and your professional responsibilities, this feels like the right decision to you.

Box 5-10

Ethically Right for Me? II

We discuss this question in Box 2-7, when we talk about Kelly's problem in meeting the state environmental law while still harming the environment. But you may think that this situation is different. Kelly knew what the law required, and she knew that her firm, Allegheny, while technically in compliance, was planning to cause harm to the environment. And this was a flagrant act because Kelly's boss didn't try to pretend otherwise. In Chris's situation, the professional obligation isn't clear and personal interpretation is required. Are there some situations in which quite different decisions could be equally right, depending on the individual interpretation—and the circumstances, of course?

Suppose we apply DeGeorge's criteria to this situation. Chris has certainly exhausted all available options within the company and has documentation to show that there is a problem. Chris is also reasonably sure that if the press got this story, something would be done to prevent a tragedy. The only question is personal cost. What will it cost to be a moral hero?

Ethicists argue that there is a limit to how much we are expected to sacrifice in order to do the right thing. For example, the switchman in Nazi Germany who pulled the levers that sent the trains to the extermination camps knew very well what was happening. Should he have protested? Certainly his loss would have been great, and few of us would argue that he had a moral responsibility to challenge the Nazi government.

In Chris's case, the cost will most likely be getting fired and having some difficulty finding a comparable job in structural engineering. Since this has been Chris's life's work, it will be a heavy price to pay. On the other hand, Alex has a good job, and an engineering education can be used in many ways. Maybe this will allow Chris a chance to develop new and challenging roles.

You meet with Sarah and Joe to tell them you want to be reassigned.

Sarah is furious with you, but, to your surprise, Joe accepts your decision calmly. "Figured you'd probably say that. Yesterday called Ken, who's supposed to start on the Fort Myers apartment building today. Showed him the hotel plans, and your stuff. He doesn't think the risks are nearly as great as you do. There is always a risk when you build anything, and in this case he feels that we are well within the range of acceptable risk. He's also confident he can still finish the building on time. Ken's not quite in your league, but he's very experienced and I'll go with his judgment."

"OK," you say with relief. You feel like a heavy burden has been lifted from your shoulders. This is no longer your concern.

"And by the way," Joe continues. "No reason to tell Timmo anything about this. They've no reason to suspect anything and telling them will just complicate matters. Agreed?"

> **Box 5-11**
>
> ## Trusting the Experts
>
> Remember that even though he runs a successful engineering consulting firm, Joe is not an engineer. Like the senior author (ethicist Alastair), he is a person who is trained in the humanities, is fascinated by engineering, and believes everything that engineers tell him. You'll recall that Joe originally went into business with his engineer wife Brenda, and that he is used to relying on engineers—particularly Sarah—for technical expertise. In the Asmara case, Joe receives conflicting advice. Sarah disagrees with Chris and does not believe the overhang is unsafe, and she is senior. Even if Chris is right and the overhang is in fact unsafe, Joe has no reason to prefer this view to Sarah's.
>
> An engineer might say that since Joe isn't an engineer, he really doesn't have the expertise to make the decision. But those who do have the expertise are giving him conflicting advice.

> **Box 5-12**
>
> ## Deception II
>
> Once again, we have the question of whether it is wrong to withhold information. The information that there have been design mistakes is certainly being withheld from Timmo. However, if the mistakes are rectified and there is no risk or cost to Timmo, is there still an obligation to inform them?
>
> Withholding information is often a form of deception, especially when the deceived persons would feel that they were entitled to know that information. Is deception just as bad as telling a lie? Consider this case:
>
> Helen's Travel Agency advertises very attractively priced package holidays to Phuket, Thailand, available in October only. The window display includes posters of shiny, happy, sun-tanned people having a great time swimming, surfing, sunbathing, and dancing under the stars. You purchase a holiday for you and your family. The trip is not a success, however, there is constant rain and thunderstorms throughout your ten-day holiday. A battered copy of the *Lonely Planet Guide to Thailand* that you read in a restaurant informs you that Phuket averages 22 rainy days in October.
>
> You might well feel that Helen's ads are misleading even though no false claims were made. Helen would probably defend her ads by saying that she wasn't practicing deception at all. You chose to interpret the posters as promising a sunny holiday; you could have asked her about the October weather in Phuket, or checked a travel book: Shouldn't you have wondered why the holidays were so inexpensive?
>
> Deception is not like solitaire; deception requires another person, and this other person has to be an active participant in the deception. One could then argue that it is the responsibility of the other person to prevent the deception by asking the right questions ("How come these tickets are so cheap?"), but sometimes this is not possible. Journalists, for example, writing in newspapers can deceive by withholding important information, and the reader most often has no chance to ask the questions that would prevent the deception. This is much the same in engineering. Engineers can deceive their clients, and often the clients have no way of knowing what questions ought to be asked in order to prevent such deception, placing a high ethical burden on the engineer to be truthful and forthright.

You nod.

"What else you got on your plate?" asks Joe.

"Now, nothing pressing," you admit.

"In that case, why don't you take over the Fort Myers job?" asks Joe.

"Sure," you say with relief and a renewed confidence.

"Fine. You'll need to spend a lot of time in Fort Myers. Contractor is fairly inexperienced. Never handled a job this big before, so you'll have to hold his hand. We advised Vito Marini, that's our client, the developer, against using him but he insisted—his brother-in-law."

"By the way, take the family for part of the time—didn't you say that Alex was taking a sabbatical to write a book? Sorry we can't give you time off for that Caribbean vacation you were planning, but some Florida winter sunshine will be a change. Vito will find you a nice apartment."

"Thanks for giving me a break on this one."

"No problem. Respect someone who has strong convictions. But remember—not a word to anyone about the hotel."

Lucky he doesn't know that you have already spoken to three people about it, you think. Of course, you can trust Alex, Shawn, and Kelly to keep it confidential.

Box 5-13

Confidentiality

Confidentiality is an especially difficult concept in professional ethics.

Because professionals are supposed to promote the interests of their clients, a strong code of confidentiality has evolved for the professions. The strength of this code varies from profession to profession, with the lawyers probably having the strictest rules against disclosure. The argument is that the lawyer can best represent a client if the client tells the lawyer the truth, and this is only feasible if the lawyer cannot be called to testify. There are times, however, when lawyers are required to reveal information divulged by their clients, such as when a life is being threatened and the information can be useful in preventing another tragedy. Physicians also have strong codes of confidentiality because patients can best be treated if they are totally truthful about their own problems. A physician will never divulge information about a patient, except when other lives are at stake. A particularly difficult situation is when a physician finds that a patient carries the HIV virus. Should this be reported to the authorities, or not? Most U.S. states have passed laws that require physicians to report such information, thus superceding the confidentiality requirement.

In engineering, the client is similarly protected, but since there is not the same potential for personal loss (of freedom or health), the rule is not that strong. Journalists would like very much to believe that confidentiality of sources supercedes their requirement to testify in court, but the courts have consistently overridden this principle. Many a reporter has spent time in jail for not revealing his or her sources.

Wednesday, November 23

You've arranged to meet with Ken to discuss the Asmara. You are determined that he will be fully informed about your views on the design.

Ken has been with the firm for 20 years. In fact, he was one of the three original structural engineers at Pines. He's not exactly innovative—some would say dull—but he's respected for his competence and loyalty to the firm.

"I see that Ali sealed the drawings," Ken observes.

"It was his first job after getting his PE license. He was very proud of it."

"Who actually did the structural analysis for the overhang?"

"Apparently it was Jerry, a young mechanical engineer we hired just a few months ago. I doubt if he had any experience with concrete, and Ali should have know that. He should have checked the calculations before he sealed the drawings."

You both again think of the tragic accident. You continue:

"Ken. I think you should run through those calculations for the overhang. I know you don't have to seal any drawings, but it would be a prudent thing to do. I'm not at all sure that the reinforcing is correctly placed, especially in the creases between the sails."

"I'll get on it. But one thing, Chris . . ." Ken says hesitatingly.

"What?"

"Not a word about this to anyone, right? This is confidential information. It stays in the firm. OK?"

"Sure, sure," you hear yourself say. "No problem."

"And this conversation never happened," Ken says as he turns and walks away.

Box 5-14

Moral Development of Engineers

How do our moral values develop? Perhaps moral values develop like other skills such as speech or physical strength and coordination. One influential analysis of such moral development was proposed by Lawrence Kohlberg (1973), who suggested that the development occurs in distinct stages. A developing child moves from stage to stage as a result of encountering moral problems for which the earlier versions of morality do not apply. Such "cognitive dissonance" results in progressive movement to higher stages, from which there is no return to a lower level.

Richard McCuen (1979) observed that such an analysis could be applied equally well to the development of engineers. Following his lead, Vesilind and Gunn (1998) suggested that the levels of engineering moral development can be described as the following:

Stage 1: Pre-professional I

The engineer is not concerned with social or professional responsibilities. Professional conduct is dictated by the gain for the individual, with no thought of how such conduct would affect the firm, the client-engineer relationship, or the profession.

Stage 2: Pre-professional II

The engineer connects conduct to marketability. While the engineer is aware of the ideas of loyalty to the firm, client confidence, and proper professional conduct, ethical behavior depends on the motive of self-advancement.

Stage 3: Professional I

The engineer puts loyalty to the firm above any other consideration. The firm dictates proper action, and the engineer is freed from further ethical considerations. The engineer concentrates on technical matters, becomes a "team player" within the firm, and ignores the ramifications of the job on society and on the environment.

Stage 4: Professional II

The engineer retains loyalty to the firm but recognizes that the firm is part of a larger profession, and that loyalty to the profession enhances the reputation of the firm and brings rewards to the engineer. Good engineering practice becomes that which helps the profession—and not necessarily society in general.

Stage 5: Principled professional I

The engineer recognizes that service to human welfare is paramount and that this brings credit to the firm and to the profession. The rules of society determine

(continued)

Box 5-14 (continued)

professional conduct. Where professional standards do not apply or are in conflict with the prevailing morals of society, society's values take precedence.

Stage 6: Principled professional II

The engineer follows rules of universal justice, fairness, and caring for fellow humans. This level is the most complex because acts of justice and caring can often contradict the prevailing social order and/or professional code of ethical conduct.

While we believe this account identifies different types of engineers, we think it is unlikely that engineers actually go through such a moral development. In fact, most engineers probably start out in the higher stages of moral behavior, and as their career develops, they drop lower on this scale as responsibilities increase. It is possible to use this scale to identify various engineers as working in the different stages of engineering morality.

Discussion Questions

5-1. What if, in your professional opinion, legal standards for environmental contamination are inadequate, as indeed they sometimes are? Suppose that your firm's perfectly legal discharge is killing fish because the state BOD (biochemical oxygen demand) limits are based on inadequate science. At some time in the future, the state will probably get around to setting ecologically sound discharge levels, if only in response to public pressure. Meanwhile, as an engineer, should you work to try to make sure that your firm complies with what you believe to be adequate environmental standards, for instance, avoiding killing fish? Or go along with current practice, while working to have standards improved? Or should you just get on with your job? How would you have responded to Kelly's predicament (Box 5-2) and why?

5-2. According to philosopher Tom Regan (Box 5-1), even if you could guarantee to kill an animal instantly and painlessly, you would still be failing to respect its right to life, which is based on the fact that as a being with a life of its own, it has inherent value. What do you think of the inherent value argument? Do you believe animals have an inherent value above and beyond their instrumental value such as food, source of power, or just enjoyment (for humans, of course)?

5-3. Do we have an obligation to take account of our environmental impacts on nonhuman living things? Is this a moral obligation? Can a case be made for moral considerability for fish and plants, or is Arthur right to dismiss them as unimportant as far as his business is concerned? (Box 5-1)

5-4. Many ethicists believe that there is no significant ethical difference between lying and deception. Sissela Bok (1978) defines deception thus: "When we undertake to deceive others intentionally, we communicate messages meant to mislead them." Lying, according to Bok, is simply a *verbal* way of deceiving people by the use of statements that you know or believe to be false. The aim and, if it is successful, the result of deception is to have the victim believe that something is the case which is not the case.

Whether you use a false statement or some other means to create this false belief is ethically immaterial. What do you think?

5-5. What other options does Chris have (see Box 5-10)? List all the options and suggest most likely outcomes if Chris chooses any of these options. Which one would you choose, and why?

5-13. If Joe, a nonengineer, should not make the decision concerning the safety of the overhang, who should? He is the president of the firm, remember, and he receives conflicting advice from his engineers. Suggest alternative ways of making this decision.

5-14. At what level of engineering moral development was Ken acting in our story? Why do you think so?

5-15. Much has been written about the Challenger disaster. Was this calamity an engineering failure or a management failure? You can read about this case in any number of books or web sites. One of the best sources of engineering ethics information is found at ethicsonline@cwr.edu.

5-16. Can one lie to a rock? Can one deceive a rock? Is there a difference?

References

Baxter, W. 1974. *People or Penguins: The Case for Optimal Pollution*. New York: Columbia University Press.

Bok, S. 1978. *Lying: Moral Choice in Public and Private Life*. New York: Pantheon Books.

Broome, T. 1999. "The Concrete Sumo." *Science and Engineering Ethics* 5 (4): 541–567.

DeGeorge, R. 1981. "Ethical Responsibilities of Engineers in Large Organizations." *Business and Professional Ethics Journal* 1 (1): 1–14.

Duncan v. Missouri Board for Architects, Professional Engineers and Land Surveyors, Missouri Court, Eastern District, Division Three, 1988, 744 S. W. 2nd 524. Quoted in J. Sweet. 1989. *Legal Aspects of Architecture, Engineering and the Construction Process*. St. Paul, MN: West, p. 174.

Ellin, J. 1982. "Special Professional Morality and the Duty of Veracity." *Business and Professional Ethics Journal* 1 (2): 75–90.

Frees and Nichols. 1998. *Ethical Conduct Policy*. Quoted in NIEE newsletter, May.

Gagne v. Betram, 1934 43 C, 2d 481 275 P2d 15.

Glazer, M. 1983. "Ten Whistleblowers and How They Fared." *Hastings Center Report* 13 (6): 33–41.

Kohlberg, L. 1973. *Collected Papers on Moral Development and Moral Education*. Cambridge, MA: Harvard University Press.

McCuen, R. 1979. "Ethical Dimensions of Professionalism." *Journal of Professional Activities*, ASCE, 105 (E12).

Muyskens, J. 1982. *Moral Problems in Nursing: A Philosopher's Perspective.* Totowa, NJ: Rowman and Littlefield.

Regan, T. 1983. *The Case for Animal Rights.* Berkeley: University of California Press.

Singer, P. 1975. *Animal Liberation.* New York: Avon Books.

Taylor, P. 1986. *Respect for Nature: A Theory of Environmental Ethics.* Princeton, NJ: Princeton University Press.

Vesilind, P. A., and A. S. Gunn. 1998. *Engineering, Ethics, and the Environment.* New York: Cambridge University Press.

Werhane, P. 1988. "Engineers and Management: The Challenge of the Challenger Incident." *Journal of Business Ethics* 25:319–340.

6

Professional development

Wednesday, November 23

It's evening, and you're relaxing at home when the phone rings. "Yo, Chris, it's Earl!"

Earl is a friend of yours. Well, he's not exactly a friend. You met him at the local engineering club ten years ago and you've spent quite a bit of time together, but you're not really close. When you met him, Earl was working as an electrical engineer for General Electric. He had other ambitions though—he was enrolled for a doctorate at State, and he wanted to have an academic career. You wondered if he was up to it, so you were very pleased for him when he received his doctorate and even more pleased when he was appointed assistant professor at Willis University. Willis has a good engineering college that specializes in electronics and electrical engineering. At that time, he had recently been divorced and remarried to a woman some ten years his junior. He'd taken a huge salary loss from his previous job—his career with General Electric had been very successful, and he'd several times told you that he'd been headhunted by other large national companies—but his new wife, Alison, had a successful career in publishing. In fact, Alison had been very supportive of his career change.

"Earl, good to hear from you, what's up?"

"Oh, this and that. Wondered if we could maybe have lunch some time, haven't seen you for a while."

"How about Monday?"

"Sure."

Monday, November 28

Earl likes Italian food, so you arranged to meet at Adriana's. When you arrive, he is already seated at a table, chewing on a hunk of Italian bread. He's always been a big guy—well, a fat guy—but he must have put on 20 pounds since you last saw him and you're reminded of Stubby Kaye.

"Chris, how ya doin'? I already ordered the special, linguini and clams, OK?"

"So, how's things at Willis?"

"Hey, let's not talk about that. You see *The Fifth Element*? I just rented it from Blockbuster. Whaddya reckon it's most like, *Star Wars, Alien, Blade Runner, The Matrix, The Hitchhiker's Guide to the Galaxy,* that Spice Girls video, what?"

"All of the above, I guess. Didn't you just love the scene where . . ."

But then the linguini arrives, and Earl doesn't have much to say for a while.

You're halfway through your lunch when Earl, who has finished his, says, "Really, I got a problem."

"What's that?"

He's silent for a while, staring straight ahead. Finally he says, "I didn't get tenure. The review was negative. I appealed, but they turned me down. I haven't told Alison yet."

Box 6-1

Tenure in Engineering Schools

Academic freedom is absolutely essential to education. Academic freedom is not a license to do or say anything you want, but academic freedom does protect professors from getting fired for saying things that might be unpopular with the administration or politicians who, at least in the case of public universities, have the final say in university affairs. Academic freedom, so the argument goes, is only possible if the university cannot fire professors for speaking out on issues of importance and for speaking the truth as they see it. The contract between faculty and university must, therefore, contain a "no fire" clause. This is called *tenure*.

Tenure as we now know it was originally established at the University of Wisconsin in the early 1900s. Wisconsin was a hotbed of progressive ideas, and professors speaking radical thoughts believed they needed the protection of the university. The fledgling American Association of University Professors (AAUP), formed in 1915 by a number of prominent faculty from 60 institutions, quickly became involved in the defense of academic freedom. Its first published work was the 1915 Declaration of the Committee on Academic Freedom and Tenure. In the 1940s, AAUP, the closest organization that university faculty have to a union, promulgated a set of guidelines that defined what tenure at universities was to mean. This AAUP document is still the operative statement on tenure and is respected by most universities even though the document has no legal standing.

Competitive pressures for hiring the best professors caused most universities to accept the AAUP definition voluntarily and adopt a system for tenuring faculty, a system that has changed little over the past 60 years. Basically, the university agrees to provide a job for the faculty as long as the members adhere to a code of conduct that includes doing their job (teaching classes) and being morally upright. Morality used to mean exclusively sexual morality, but in the last few years professors have lost their jobs for social infractions such as getting caught committing petty crimes. Universities actually have a number of ways of dismissing tenured professors; the most drastic being simply declaring that all tenure is revoked and cleaning house. Revocation of tenure when the department is eliminated has happened in a number of places, and there is little the affected faculty can do. James Madison University, for example, closed its graduate physics program and revoked tenure, and Bennington College has eliminated tenure for all faculty. Tenure therefore should not be thought of as a lifetime contract as much as a simple agreement of goodwill between the faculty and the university.

But today tenure is under fire, and its advantages are being hotly debated. Three arguments are commonly presented against tenure. The first argument is that tenure does not accomplish its aims—to guarantee academic freedom. AAUP claims that the absence of tenure will create chaos in academia, and yet we have institutions where the rights of faculty are strictly limited on religious or other grounds while other institutions have either abolished or severely curtailed tenure with no apparent ill effects. In support of this argument, note that all of our untenured faculty members do not, by definition, have the protection of tenure, and many of them are socially and politically active. Although the data would be impossible to obtain, it is unlikely that a single nontenured faculty member has been dismissed in the past 50 years for making unpopular social or political statements. The coupling of academic freedom with tenure is therefore

(continued)

Box 6-1 (continued)

inappropriate. In today's social climate, there is no clear cause and effect. Academic freedom does not exist because of the tenure system.

The second argument against tenure is that it does not produce the best faculty. There is no doubt that some of our best teachers are denied tenure because they are not sufficiently productive in research or scholarship. For recent Ph.D. graduates, the thought of up-or-out tenure decisions is not an inducement when they are seeking their first jobs. Many graduate students are pleased to find that some universities do not have strict tenure policies or that non-tenure-track appointments are possible. These students have a sense of their own worth and would want to prove their value to the university unencumbered by the tenure decision. They shy away from universities that have strict tenure policies and a poor record of retainment. Tenure is therefore unlikely to enhance the quality of an academic institution.

Finally, tenure is and has been used by lazy or unethical senior faculty to provide a secure income while they pursue outside interests. Although some outside work can be useful if brought to the classroom or research laboratory, often the interests have little to do with their scholarly field. One senior professor at the University of North Carolina, for example, runs a funeral home and comes to the university only once a week to teach a small seminar that has not changed in 20 years, all the time drawing a full professor's salary.

Why, then, does tenure exist? Many faculty deeply and firmly believe that academic tenure is absolutely necessary to retain their academic freedom. DeGeorge, in his book supporting tenure, states flatly: "The justification (for tenure) is that academic tenure is the best means our society has devised to secure and preserve academic freedom" (DeGeorge 1997).

Without doubt, academic freedom is important to society because in the free university is where truth is to be discovered. As U.S. politician Adlai Stevenson (1900–1965) once remarked, "A free society is a place where it's safe to be unpopular." If academic tenure is the best way to preserve academic freedom, then academic tenure is good for society and should be preserved.

But this argument takes a huge jump in logic. If we agree that academic freedom is good for society and should be preserved, how can we show that the present tenure system is the only way to guarantee academic freedom? We can prove that this is not causal by proving either that such a relationship does not exist or that there can be other equally feasible solutions. We cannot do the former, but is it possible to show that academic freedom can be preserved and guaranteed by means other than tenure? Of course. Take, for example, the situation in which all faculty have civil service jobs (as is the case in many countries, such as Canada) and firing civil service employees is notoriously difficult. Civil service employees are protected by the First Amendment to the U.S. Constitution, guaranteeing free speech. This amendment says that the government (the employer in this case) cannot prevent people from speaking their mind, and thus academic freedom would be guaranteed. In New Zealand, where all the universities are public, there is no tenure in the American sense. All new appointments, after a four-year probationary period, are awarded a "continuing appointment," which assures the professor of a job unless the university can prove unsatisfactory performance (which the employer must endeavor to assist the employee to rectify), misconduct, or incapacity, or unless the position becomes redundant (in which case compensation must be paid). This system allows for job security for the vast majority of professors while allowing the universities to get rid of truly objectionable professors.

We are not advocating that all faculty become civil servants or move to New Zealand, but we are suggesting that DeGeorge's argument for the retention of tenure in universities is a hollow argument (Vesilind 2000).

"Oh, I'm so sorry, Earl. That must be real hard. But maybe it isn't so bad. Can't you apply again next year?"

There is another silence, during which you pick up the last of the clams, which are excellent, abandoning the linguini, which isn't.

"Yeah, you'd think so, but that's not how it works. The letter I got was: no tenure, no job, so long sucker."

You remember conversations with Alex about the gut-wrenching experience of tenure applications, and the devastation that people feel when they've been turned down. And about the campaigns that disappointed tenure-track academics have run, getting their friends to write letters of support to tenure appeals committees, almost always without success.

"You must have *some* idea as to why they turned you down?"

"Yes, of course. I have been warned for several years that my level of outside funding and publication record is below average. I understand the need to do that, but this is not where my heart was. I just love teaching. I love being in the classroom and working with students one-on-one. This year, I had almost half of the senior class doing senior design projects with me. It took an immense amount of time, and the other faculty were tickled that I would do all the senior design projects so they didn't have to."

"And I am sure all the students appreciate your efforts. Engineering is still, because of people like you, the best 'liberal arts' education any young person can get."

Box 6-2

Famous Engineers in History

One of the shortest books in the world might be entitled *Famous Engineers in History*. If an otherwise intelligent person is asked to name some of the famous engineers in history, it is unlikely that he or she will be able to name many. Some may know about the Robelings (father and son, John Roebling, 1806–1869, and Washington Roebling, 1837–1926), who built the Brooklyn Bridge, or about Marc Isambard Brunel (1769–1859), a French-born British engineer who designed the first subway tunnel under the Thames River in London. But that would be about it. A very short book indeed. Engineers simply do not become famous for the work they do.

On the other hand, many engineers have become famous for contributing to society in ways other than in engineering. Engineers have become politicians, musicians, artists, soldiers, sports heroes, and more. Here is a sampling of some engineers who have "gone over to the dark side" and become famous:

Scott Adams, creator of the *Dilbert* comic strip
President Herbert Hoover, a mining engineer
President George Washington, a self-taught engineer and surveyor
President Jimmy Carter, a nuclear engineer
President Ulysses S. Grant, a graduate of the West Point engineering program
President Yassar Arafat, Palestinian leader
Chairman Leonid Brezhnev, former leader of the USSR
Astronaut Neil Armstrong, the first man on the moon
Herbie Hancock, jazz musician
Lee Iacoca, former chairman of Chrysler
Alexander Calder, sculptor and artist
John Sununu, President George H. W. Bush's chief of staff and former governor of New Hampshire
Arthur Nielsen, inventor of the Nielsen ratings
Tom Landry, former coach of the Dallas Cowboys football team
Bill Koch, captain of the U.S. America's Cup yacht racing team
Alfred Hitchcock, movie producer (*Rear Window*, *Psycho*, and many more)
Roger Corman, movie director (the original *Little Shop of Horrors*) and producer

Engineers all. The list could go on and on. What part did their engineering education play in helping them become world leaders in their work? Roger Corman is explicit in crediting the engineering education (Stanford) with teaching him how to produce good stuff, on time, on a limited budget. Perhaps there is something to that. Or, as my (engineer Aarne talking here) wife's grandmother once advised her: "Never marry a man who does not carry a pocket knife."

"Yeah, and it was fun. But it all cut into my efforts to get research funding. At some engineering schools, the expected level is $500,000 per year now. It's not quite that bad at Willis, but getting there. They have "MIT-syndrome" just like so many other engineering schools."

"Not getting tenure must be so dreadful, I really sympathize," you say, inadequately. "What other opportunities do you have?"

"I'm fifty years old, Chris. I took a chance on an academic career, and I didn't make it. Do I look like a hiring opportunity for anyone?"

You have to admit that he doesn't. Still, smiling brightly, you say, "Earl, you have to be positive. There are lots of opportunities out there. There's something for you, I'm sure!"

"I've been looking and I haven't found any. Say, Chris, I was wondering . . ."

"Uh, I'm not sure what the situation at Pines is, maybe I can . . ."

"No, no, no, no, no! I'm not hitting on you for a job. But, uh, I'm desperate. I applied to Pulson University. I understand they are looking for new faculty, . . . and I heard Alex's Uncle Max is provost there."

Uncle Max is very fond of Alex and he likes you a lot, too. If Earl survives the search committee and the decision goes to the provost, a plug for Earl from you could help his chances considerably. "I'd be glad to write a letter of recommendation for you, but I know very little of your work. Never did understand what you were doing in your dissertation. I am afraid my letter would not be of much value."

"Well, . . . maybe, uh, I wondered if you could talk to him personally, maybe."

"Well, I could think about that. I'll have to talk to Alex, of course."

Box 6-3

Networking

Networking is the process of getting to know people who might, in the future, be of some service to you. When Earl asked Chris to have Alex's Uncle Max intervene on his behalf, he was using a network.

Networking, known in many English-influenced countries as the Old Boy network, has attained a negative image as the system by which undeserving people, by virtue of their connections (family or professional), can gain an advantage over more deserving competitors. There is no doubt that this occurs, and that it is immoral because the connections are not available to all the participants. Suppose a student applied for a job, and he has an uncle already working in the firm. This is a good connection, and the chances are good that the student will get the job even though he might not be the best applicant. The company does favors for its own people. I scratch your back, you scratch mine, is the motto. The person harmed by this is the more qualified applicant who did not get the job.

On the other hand, networking is not intrinsically immoral. Suppose a student applies for a job, and the personnel director calls one of the professors for a reference. The professor gives her best estimate of the qualifications of the student, and the personnel director uses this information to make a decision. This is, we all agree, a perfectly agreeable process.

But now suppose the personnel director knows the professor personally. He calls up and says, "Martha. How are you? How goes the research?" and then asks her about this student. Is the information the personnel director gets from the professor worth more or less because he knows her personally? Certainly more, because he knows the quality of the person giving the recommendation. If he does not trust the opinion of a professor, he would never call him or her. He knows Martha will be honest with him. If Martha either says something that is not true or misjudges the student's abilities, the personnel director will not ask her again, and Martha knows this.

Networking is therefore a useful process for getting to know who can do what. Yes, often it is abused, but the process itself is a useful and ethical one.

Discussion Questions

6-1. Suppose you decide to get an advanced degree in engineering and that you want to be a professor. When you receive your degree, you have a decision to make about what kind of college or university to work for. One of the main variables is whether or not the institution has tenure, and how tenure is awarded. Knowing what you now know about tenure, would you choose to work at a university with or without tenure for its faculty, everything else being equal? Why?

6-2. Tenure in most American universities, in engineering schools, is given on the basis of research performed and published, and research funds obtained. Many engineering professors disagree with this narrow requirement for tenure because neither engineering skills nor teaching excellence is rewarded. Others believe that engineering science, the performance and publication of research, is and should be the sole criterion for tenure. What do you believe? Write a statement outlining the skills and qualifications for the kind of professor you would want to have tenured at your university.

6-3. Would it be ethical for Chris to try to influence Uncle Max to hire Earl? Why or why not?

6-4. In Box 6-1 we present a number of arguments for and against the tenure system. What do you think? Is it a good system? Is it in the interest of students? Should tenure depend largely on research output or on teaching? Why?

6-5. Do a structured interview of the faculty at your university. Ask at least 20 faculty members if they support or oppose the tenure system as it presently exists at their university. Write a report on their responses.

6-6. When you apply for a job, you will be asked for references. Who would you choose for your three references, and why would you choose these three persons?

6-7. Describe a situation in which you believe networking has been to your advantage or disadvantage. Were any moral rules broken in this incident?

References

DeGeorge, R. T. 1997. *Academic Freedom and Tenure*. Lanham, MD: Rowan and Littlefield.

Vesilind, P. A. 2000. *So You Want to Be a Professor?* Thousand Oaks, CA: Sage.

7

Solicit or accept gratuities

Sunday, November 6, to Friday, March 12

When you arrive, you find the Fort Myers project in shambles, but you straighten things out. The winter flows by and it's already summer in Florida. Antony, the owner of the contracting firm, is certainly inexperienced. But he is keen to learn, and you enjoy passing on your knowledge and helping him deal with subcontractors. You even help him out with a redneck city official, Kurt, who thinks all Italian Americans are crooks and who constantly disrupts the work with pointless snap inspections. He backs off when you drop hints about Vito's mob connections who don't like people like Kurt. This isn't true, so far as you know, but you feel fully justified in deceiving him for a good cause. You reason that Kurt doesn't deserve respect if he's not prepared to treat other people decently, and anyway, it's a harmless enough deception.

Box 7-1

Deception III

Is deception as wrong as lying? Is it even wrong at all? The eighteenth-century German philosopher Immanuel Kant (1724–1804) believed that lying is always wrong, even if it is the only way to save a life; perhaps because his position on lying was so extreme, he believed that deception in itself was not wrong. He gives this example: Suppose you wish your neighbors to believe that you are taking a vacation, so you ostentatiously pack your carriage. You are not lying to your neighbors: It is up to them to interpret your behavior as they choose.

We could say: It takes only one person to lie, but it takes two to deceive, because deception can't happen unless there's someone to interpret words or actions in a particular way. Ethicist Joseph Ellin, in a discussion of medical ethics, defends what he calls "therapeutic deception"—deception that is practiced for the benefit of the patient, because the physician believes that the patient may be better off not knowing the truth about his or her condition (1982). Lying is wrong because it violates the patient's trust by making a false statement, whereas a person who is being deceived is a "participant," a party to the deception—he or she chooses to interpret an ambiguous statement in a particular way. Thus, Helen (the travel agent who sent you to Thailand during the rainy season in Box 5-12) did not lie to you: You chose to assume that the posters in Helen's window display accurately depicted Phuket in October, as indeed it did for nine October days. In this view, deception is not necessarily wrong, or at least not as wrong as lying.

Other writers argue that deception—or even lying—is not wrong if it is harmless, especially if everyone is

(continued)

> **Box 7-1** *(continued)*
>
>
>
> doing it. According to Scott Adams of *Dilbert* fame, everyone buys the *Economist*, but no one reads it. The point of buying the magazine, and leaving it around for everyone else to see, is to create the impression that you are a very smart person.
>
> Consider this related case: In a large course at a university, teaching assistants (TAs) are hired to grade various objective tests such as multiple-choice quizzes. As the professor in charge of the course, you always check a random selection of the work before returning the results to the students, in case the assistants have made errors. On one occasion, you discover that one of the TAs has made numerous errors in grading the first assignment for the course. You further find that there is a pattern to the errors that systematically works against students with "ethnic" and "foreign" names, and advantages students with Anglo-Saxon names. Inquiries reveal that the TA maintains an Internet page devoted to a racist hate group. You confront the TA with all this information, and she admits her misdeeds. Since her contract of employment specifically requires her to grade all work strictly on its merits, you summarily dismiss her and regrade the work accurately. Thus, the students have all received the correct grades for their work. Now that you have fired the TA, thus ensuring that the problem won't occur again, do you have an obligation to inform the students of the (now corrected) misgrading?
>
> Is the professor in this case essentially in the same position as the responsible engineer on a project who fixes up technical errors? In the case of the students, they did not need to know that their grades had been corrected. Similarly, the client would not need to know that during the design process, mistakes made by engineers had been caught and corrected. Presumably, in both cases the professors and the engineers would admit to the original mistakes and the corrections if questioned, but it would not be in the students' or the client's interest to ask such questions.
>
> In a typical engineering office, there are people whose most valuable asset is their ability to check drawings. They will go over every re-bar, every bolt, and every pin, to make sure it is right. On my (engineer Aarne talking here) first job at an engineering office, I had one of my drawings checked by such a person, who predictably found a lot of errors. I was upset and mad and angry, until I realized how grateful I should be to the checker. The mistakes were found *before* the drawings left the office, not afterwards. I would have been a great deal more upset if the facility had actually been built and *then* the mistakes were discovered.
>
> Should there have been a person like that at Pines who would have caught the mistakes made in the design of the hotel overhang? Presumably, but the design was so complex and so unusual that it could easily have gotten past the "re-bar guy." In the end, the design is still the responsibility of the engineer who seals the drawings, testifying that they are accurate and that the structure will be safe.

You and Vito become good friends, and on several occasions he invites you over to his home for dinner.

As a result of your excellent work, the building is ready for occupation in late February—four weeks ahead of schedule. Vito is delighted! He wasn't counting on being open for business until after the tourist season. Antony is delighted too because he made such a nice profit on the job. Vito is so pleased with his work that he's proposed a partnership, which will set Antony up for the future. And Joe is delighted because Vito wants Pines to do all their design work.

Saturday, March 13

You receive a call from Vito, inviting you and your family to the launch of Marini Properties, the new company that he and Antony have established.

To your surprise and pleasure, you're the guest of honor. Vito makes a speech in which he praises you for your fine work and assistance. "Chris, you did such a great job. We'd have been lost without your expertise and willingness to pitch in and help. So Antony and I would like to make you a little gift, to show our appreciation."

He signals to two attendants who whisk away a large screen from the podium—to reveal a sparkling new silver Alfa Romeo Spider! Everyone bursts into applause and enthusiastically drinks a toast to your health. Vito and Antony hug you. Your family is thrilled. Your 6-year-old son, James, rushes over to the car, climbs into the driver's seat and, to general delight, sits behind the wheel and goes "Vroom! Vroom!"

Alex gives you a huge grin. "Now you can get rid of the grungemobile, darling. Isn't it great!"

"Sure is," you say.

Inside, though, you're troubled. Pines has a strict policy that employees are not allowed to accept gifts worth more than $100 from clients.

Box 7-2

Corporate Gift Policies

Some organizations allow employees to receive gifts from firms and organizations with whom they do business. In effect, they leave it up to employees to decide what is right for them. One of the authors (ethicist Alastair) has two friends who work for firms that have quite different policies. One, whom we will call Michelle, is purchasing director at a small consulting firm with a turnover of about $10 million, and she is responsible for about $1 million of equipment and supplies annually. Her policy is that she will not accept *any* gifts, however small. She has informed all the firm's suppliers of this, and she always returns gifts. The other friend, whom we'll call Jerry, is human resources manager at a somewhat larger company with a turnover of some $40 million. He does not handle a large budget but frequently makes decisions that involve considerable sums of money. He always accepts gifts. Mostly, these are presented at Christmas by secretarial and employment agencies and suppliers of software with whom his firm has done business. The gifts have included gift baskets containing gourmet foods and luxury liquor (which he shares with his staff), and corporate hospitality at major sporting events, which he personally attends. He makes no attempt to hide the fact that he receives these gifts, and no one has ever suggested that they have ever affected his judgment. Indeed, it is hard to see how they could, because the market for the products and services he deals with is extremely competitive and many companies give him gifts, including those with whom he does not do business. Moreover, his firm has a highly efficient internal auditor who would probably be aware if he was unduly favoring a particular supplier.

Other companies allow gifts to be received but only up to a certain value. This allows a supplier to show its appreciation with a small gift but not to the extent that it might influence the employee's judgment. The giving and receiving of gifts is an important social ritual, especially at holiday periods, and most people would accept that modest gifts such as tickets to a regular season sporting event, a bottle of wine, or an invitation to lunch are ethically acceptable.

Other organizations (including, until recently, the U.S. federal government) forbid employees to accept gifts. For instance, the University of Virginia requires that

> University employees must not accept personal gifts of any kind, including food and beverages, travel, and tickets to sporting and cultural events, from firms with which the University does business.

The reason for this draconian policy, presumably, is not just to guard against corruption, but to protect the university's reputation against public misperceptions

(continued)

86 CHAPTER 7

> *Box 7-2 (continued)*
>
>
>
> and suspicions—even ill-founded suspicions. Of course, no reasonable person would believe that a purchasing manager would give favorable treatment to a supplier who has sent him or her a bottle of wine at Christmas. The University of Virginia policymakers have assumed, however, that reputation includes the opinions of laypeople who are not fully reasonable and do not know all the material facts. Indeed, the "reasonable man" test in common law is not really about reasonableness at all. As the eminent British jurist Sir Patrick (later Lord) Devlin (1965) once remarked, "The reasonable man is not expected to reason about anything and his judgement may be largely a matter of feeling" (p. 15).
>
> When is something a gift? Michelle accepts invitations to lunch with potential suppliers but not to expensive restaurants, and only when there is a clear proposal to discuss. Jerry places no such limits on meal invitations, though he tries to ensure that business matters are discussed. He reasons that he is likely to learn valuable information especially if, as sometimes happens, the host's judgment happens to become clouded during the course of the meal (which his never is).
>
> Again, a company that fears its senior management may be influenced by a free lunch has obviously hired the wrong person. But what about larger-scale efforts? Specialist physicians are frequently invited by pharmaceutical companies to all-expenses-paid conferences in overseas locations. Often, they and their partners are flown out in business class, accommodated in expensive hotels, and invited to stay for several days after the conference (which may last for only one or two days). Usually, visits to cultural, sporting, and other events, and travel to scenic attractions, are provided. This is attractive to physicians in New Zealand (where Alastair lives) because of its geographical location. One acquaintance of his, whom we'll call Glenn, was offered a ten-day visit to the United States; the only obligation was to attend two half-day "conferences," which he assumed would be commercial presentations, in New Orleans and Las Vegas. Vouchers for gambling chips, show tickets, and visits to the Vieux Carré and a luxury cruise on the Gulf were included in the package. Glenn declined the offer on ethical grounds. He believes that such jaunts simply increase the price of the product, that he would have learned nothing useful from the presentations, and that he could not justify being paid by his employer while he was on a luxury vacation.
>
> The professional justification for attending these conferences—in the case of hospital-employed physicians, on salary—is of course that it is an opportunity to meet with colleagues from around the world and keep up with new developments. The motives of the pharmaceutical companies, however, are obviously commercial. The companies that offer the most frequent and alluring packages are those that produce drugs for common complaints such as depressive illness, asthma, diabetes, arthritis, heart disease and cancer, for which huge volumes of drugs are prescribed and where there is fierce competition between a few giant companies.
>
> Glenn, a psychiatrist who works exclusively in the public health system, makes a point of accepting only invitations to genuine conferences where he expects to learn something of benefit to his professional practice, and also of rotating the companies that invite him so he never receives more hospitality from one company than from another. Other physicians, such as his colleague Hank, he says, take several trips a year on the basis of the attraction of the location and the lavishness of the hospitality, and because the public health system has difficulty in recruiting and retaining senior clinicians, their employers usually go along with it.

But it would be insulting in the extreme to refuse the car, and you'd feel a complete heel. Your family is so pleased for you, the kids are looking forward to driving around in the cool car, and you can see Alex gazing covetously at it. Alex's old Challenger really is on its last legs.

Later the same day

Back at the motel, the kids are in bed, and you tell Alex about Pines' policy. You tentatively suggest that you might sell the car and donate the proceeds to charity.

"It must be worth at least $30,000."

"The sticker price is just over $40,000, " Alex says, "and that's without the extras: That car is loaded. Chris, you earned it. The firm won't ever know; you can just say you came into some money and bought the car out of the inheritance."

"Well, maybe we could sell it and pay off the home loan then," you suggest.

"Absolutely not! We don't need the money. We can pay off the loan in a few years anyway. And what are you going to tell James and Laura? They'll be *so* disappointed. If it comes to that, so will I. Why sell yourself short just because of your firm's stupid policy anyway? It's not like you'd be accepting a bribe or anything. Antony and Vito are nice guys, and they just want to reward you for your help. Accept their generosity—it's their culture! What if you sold it and they found out—they'd be so insulted that you spurned their gift, Pines can kiss goodbye to any more work from them! You owe it to them, to Pines, and to yourself—keep the car and enjoy!"

Monday, May 21

All things considered, it certainly seemed to be in everyone's best interests for you to keep the car. But you also believe it would be wrong to deceive Joe, who has been very good to you. On your first day back at Pines, you leave the Spider at home, drive to work in your Camaro, and arrange for a brief meeting with Joe. He's very understanding.

"Brought in the policy to protect the company's interests. Can't afford to have a reputation for accepting kickbacks—illegal anyway. A case a few years before you came to us where one of our engineers (no longer with us, of course!) accepted a gift from a supplier to recommend that we use their equipment. Turned out to be defective. Cost us a bundle and that's why I brought in the policy. But in this case, you already completed the building and brought us more business. Keep the car; treat it as a bonus. But let's keep this to ourselves."

Everything seems to be working out. Your colleagues accept your story about the bequest from Aunt Edith in Seattle, and you get tired of jokes about every cloud having a silver lining.

Discussion Questions

7-1. The justification for deception often encountered is that "everyone is doing it." Under what circumstances would this rationalization be acceptable? That is, when would deception be morally condoned if "everyone is doing it"?

7-2. Have you ever had a TA or professor deceive you? In what way? Do you think the deception was justified?

7-3. Suppose Hank (see Box 7-2) is called before the judicial review board for New Zealand physicians. What justifications would he submit for his actions? Write a one-page brief outlining his defense.

7-4. Is it fair for Joe to allow Chris to keep the car, since it is in such flagrant violation of company policy, and other employees may well have turned down gifts for that very reason? Isn't Chris receiving special treatment—compounded by the fact that Joe is covering up for him by going along with the lie about Aunt Edith? Consider the possible consequences of Joe's decision. List the consequences that could occur, the people affected, and the benefit or harm that might result. All things considered, what ought Joe to have done?

References

Ellin, J. 1982. "Special Professional Morality and the Duty of Veracity." *Business and Professional Ethics Journal* 1 (2): 75–90.

Devlin, Sir Patrick. 1965. *The Enforcement of Morals*. Oxford: Oxford University Press.

8

Self-laudatory language

Monday, May 28

Ken completes the Asmara Philadelphia on April 26—ten days late, but there was bad weather and a major subcontractor went bankrupt, so Timmo is happy and it looks like they'll put more work Pines' way. Joe, Sarah, and Ken are invited to the grand opening, and they return accompanied by Girish, a staffer from the Genalan Ministry of Construction, who expresses an interest in looking around Pines' office. Genala is a poor but rapidly developing country in Asia. You've never been there, but you worked in Singapore once and visited several nearby countries, where you developed an interest in the cultures of the region. Girish meets with all the senior engineers, and you and he get along particularly well. When it is time for him to leave, you drive him to the airport, and on the way he mentions to you that you are the only engineer he's met in America who has any real understanding of Asian cultures.

After the completion of the hotel and the convention of architects, Pines Engineering Design receives many compliments and even a few awards for its role in the construction of a most impressive building. Joe, smelling more work and money, calls you into his office.

"Chris, gotta use this for competitive advantage. Make sure our role in this project is well advertised. Thinking of full-page ads in the *Engineering News Record* and *Civil Engineering*,[1] telling the world how brilliant we are. Here, I've done some preliminary sketches as to how the ads might look." He hands you a sketch.

Box 8-1

Advertising I

This is the kind of ad Joe had in mind:

ANOTHER PINES ENGINEERING
DESIGN SUCCESS

Pines Engineering Design, the premier structural design firm in the United States, due to its excellent engineering skills and management organization, was able to work with Timmo Construction to complete the Asmara Philadelphia Hotel on time and under budget. We look forward to working on YOUR next construction project.

When you need the best, hire the best—hire Pines Engineering Design.

Apart from the claim that Pines is the "premier structural design firm in the United States," this ad is truthful enough. Construction companies advertise themselves in similar terms and no one cares. Why should engineering firms not be allowed to do the same?

---

[1] Respectively, the magazines of the construction industry and of the American Society of Civil Engineers (ASCE).

"This is very nice, Joe, but we can't use this ad. Its unethical."

"What? Are we saying anything that's not true? Is there anything misleading in the ad?"

"No, but that's not it. The ASCE Code of Ethics is very explicit about such advertising. I forget the exact wording, but it's something like 'engineers shall not advertise their work in a self-laudatory manner,' and this certainly is self-laudatory."[2]

"Sure it is. We are *good*, and we want the world to know it."

"The idea, Joe, is that if we are really good, the world will eventually know it because of our great work, not because of our advertising campaigns."

Joe is silent, considering his options.

"Well, I'm not a PE and I can run this ad if I want to!"

"If you do, you'll be in trouble with the state licensing board and the firm—which has the name engineering in its title, don't forget—may be fined. I really think you should forget the whole idea."

Joe sits down at his desk, and a look of disappointment comes over him. But he is not giving up.

"Every other profession can use self-laudatory language in advertising. Look at the lawyers, for heavens sake."

"No, actually, if you read their ads, they never say they are good, or that they can make you rich. They are very careful to only describe their services, and to say if you need a lawyer, to call them."

"Oh, OK. So it was a dumb idea. But it would have been a great ad, don't you think?"

Box 8-2

Advertising II

At the present time, all engineering codes of ethics either implicitly or explicitly prohibit self-laudatory advertising. For example, the Accreditation Board for Engineering and Technology (ABET) Code of Ethics states:

5.g. Engineers may advertise professional services only as a means of identification and shall be limited only to the following:

g.1 Professional cards and listing in recognized and dignified publications, provided they are consistent in size and are in a section of the publication regularly devoted to such professional cards and listings . . .

g.2. Listing in the classified section of telephone directories, limited to name, address, telephone number and specialties in which the firm is qualified without resorting to special or bold type.

What may be the purpose of such restrictions? First, it would certainly save the engineering firms money if they didn't have to advertise. In effect, this might be interpreted as collusion in the constraint of trade. Second, by limiting the type and size of the advertising, the engineering profession is hoping that the public will recognize that this is a "dignified" profession and that this would enhance the reputation (and the salaries?) of engineers. But the most often used argument is that these restrictions are in the public interest. If engineers are chosen only on the basis of their qualifications and not on their advertising budget, then the public should benefit because the really bad engineers would soon go out of business. Whether advertising would actually be detrimental to the general welfare of the public remains to be demonstrated.

---

[2] The actual language in the ASCE Code of Ethics is positive: "Engineers may advertise professional services in a way that does not contain self-laudatory or misleading language or (is not) in any other manner derogatory to the dignity of the profession."

Discussion Questions

8-1. Rewrite Joe's ad in such as way as to not be self-laudatory. Discuss how the new ad differs from the one Joe was proposing.

8-2. Photocopy from any telephone book some ads by legal firms. If engineering firms advertised like this, would they be considered unethical? Why or why not?

9

Contributions in order to secure work

Monday, June 29

Joe calls you into his office with the news that Pines had received an invitation from the Genalan Ministry of Construction to submit a RFP (request for proposal) for the design of a resort complex in an area that is being developed for international tourism.

Joe is ecstatic. "Great opportunity for Pines! Got Carmen to check out the Genalan economy, and their current development plan is emphasizing upscale tourism. This hotel complex is very important to them. Be a boost to the economy and create lots of local jobs. And it's a great break for us—been thinking of expanding internationally, especially after the Asmara won all those awards. Terrific opportunity for us. Want to do the response to the RFP?"

"I'd love to do it!"

"Great! Now, make it look good, but keep the bid low. I only want to break even on this—we need the experience as well as the exposure."

Box 9-1

Competitive Bidding

As professionals, engineers have long believed that it is unethical to bid competitively against each other. They argue that one would not seek out the cheapest physician if one's appendix needed removing, so why should a client seek out the cheapest engineer for a technical task? If engineers start to undercut each other through competitive bidding, then the clients will suffer because the quality of engineering will be substandard.

The U.S. Department of Justice did not see it this way, however, and called the prohibition against competitive bidding "restraint of trade." In 1972, the Department of Justice filed antitrust action against ASCE, claiming that Article 3 of its Code of Ethics, which defines competitive bidding as unethical, violates the Sherman Antitrust Act. This action occurred even in spite of the fact that ASCE had the year before, in anticipation of this action, struck Article 3 from its Code of Ethics and left the decision about whether or not to participate in competitive bidding up to the engineer. ASCE and most of its members argued that when public works are designed by the low bidder, the quality will suffer and public good will be adversely affected. The Department of Justice, however, held that if there is no market constraint on the cost of engineering design, the profession is in effect a monopoly.

After much discussion, ASCE agreed to a consent degree that forbade ASCE from including any refer-

(continued)

> **Box 9-1** *(continued)*
>
>
>
> ence prohibiting competitive bidding in its Code of Ethics or other policy documents. ASCE, however, got in the last word by adding a footnote to the 1972 version of the Code:
>
> > Under the Code of Ethics of the American Society of Civil Engineers, the submission of fee quotations for engineering services is not an unethical practice. ASCE is constrained from prohibiting or limiting this practice and such prohibition or limitation has been removed from the Code of Ethics. However, the procurement of engineering services involves consideration of factors in addition to fee, and these factors should be evaluated carefully in securing professional services.

Wednesday, July 16

To your and Joe's delight, Pines is selected as one of the three companies to make a presentation on the resort complex in Genala in September.

Today you've arranged to meet with Sarah to discuss your schedule for the next few months. She's enthusiastic about the project and wants you to drop all your current work—she'll assign it to other staff.

"Make it good," she says. "Joe wants this to work. I want you to go take a course in Genalan so we don't look like just another multinational after a quick buck. Take as much time as you need."

You have Rosemarie check around the local educational institutions, and she comes up with a total immersion course at a language school. The school promises that after completing the six-week, full-time course you'll be fluent in the language and will have a working knowledge of Genalan culture and customs. You ask her to enroll you in the course.

The course is excellent—you're good at languages and Genalan isn't too difficult. You're fascinated by the culture, and you're greatly looking forward to your trip and presentation.

Monday, September 9

You had a long and tiring flight to Ramaya, the Genalan capital, but first class on Air Genala is comfortable and you had the weekend to relax. At the newly constructed airport, you were met by an English-speaking guide, Ella, who was very impressed by your fluent Genalan.

Ramaya is rather a shock—your travels to Singapore and the Asian cultural and resort areas you've visited haven't prepared you for it. A brownish yellow haze of kerosene and car exhaust emissions fills the air, stinging your eyes. The airport is connected to the city by a new expressway, but your car passes by some very poor areas, including a landfill that, Ella explains, is home to several hundred local people. Downtown Ramaya consists largely of banks, office buildings, and luxury hotels. You feel rather guilty at living in such style amid so many poor people, but you're glad your firm will be helping to make a difference to the people's lives—if you get the contract, of course.

You've thoroughly prepared your presentation, so you're happy to accept Ella's offer to take you on a trip to an ancient temple complex on Sunday, where you spend a fascinating day.

The meeting is in a conference room at a downtown hotel. Over coffee and brightly colored sticky rice nibbles, you're introduced to about 40 government officials, developers, and journalists. Most of the guests are men—some in Western business suits, others in tropical suits or batik shirts and cotton pants, but there are also a number of women wearing expensive silk dresses.

Your presentation is obviously successful from the moment you begin your introduction (in Genalan) in which you thank your hosts for their hospitality and for the honor to be selected to present your firm's ideas on the project. You express faith in the ability of the Genalan government and private sector to develop the country and improve living standards. Actually, you're not sure what effect the project will have on the lives of any of the ordinary Genalans in the long term, but for sure there will be jobs on the project, and anyway, you're not an economist.

Joe had suggested that you make your whole presentation in Genalan, but you're not sure if you're up to it, so you switch to English. Your designs are well received, especially your incorporation of traditional themes and materials into the buildings. Pines had hired several Genalan expatriates as consultants, and you've even brought along samples of local carving and weaving for the guest bedrooms. In contrast, the other two presenters have produced rather bland designs, and neither of them can speak a word in Genalan other than "Hello."

After lunch, the review board retires to discuss the three proposals. In less than an hour, they reappear and reconvene the meeting to announce that they have decided to recommend that Pines be awarded the design contract.

Monday, September 9, evening

At the reception that evening, Girish, whom you met six months ago at the Pines office, and who turns out to be the chair of the selection committee calls you aside to suggest a walk around the hotel gardens. He is a tall, elegantly dressed Indian with an impeccable British accent.

"Chris," he says, "you made quite an impression on the review board."

"Why, thank you. Pines really wants this job."

"I think people really appreciated your approach and your understanding of our country."

"We certainly look forward to working here."

"Of course, the recommendation of the review board is not the final decision," he continues. "Ultimately it's the minister who signs the contract."

"She wasn't there today, was she?"

A mongoose-like animal springs out of the bushes and races across the lawn in pursuit of what you hope are imaginary snakes.

"No, but she still makes the decision—based on the advice she's given."

"Advice from whom?"

"From the selection committee, and from individuals . . . such as myself."

You walk around the gardens for a time, inhaling the fragrant perfume of the frangipani trees while trying simultaneously to ignore the sulfur dioxide, as you consider this. The frangipani narrowly wins out.

Girish doesn't make any further conversation; in fact, you have the impression that he's waiting for you to make the next move. You suddenly remember an *Economist* article about doing business in Genala that Alex (who actually reads it) showed you recently. The key point was the importance of personal contacts—well, you've got Girish. The article also mentioned that personal contacts usually expect personal favors, and you impulsively decide to hint at this possibility.

"I really appreciate your support, Girish, and I'm really grateful that we were invited to tender and then recommended to do the project. "

"It is nothing. We want to get the best company for the resort, and I also advised the minister to invite the other companies."

"And Pines will certainly want to show its gratitude. In fact," you hastily improvise, "the president of my company would very much like to discuss how we might work together on other projects in the future. So he asked me to invite you to visit with us again when it's convenient for you. Perhaps your wife and children would like to come too. We'd be happy to show you around. You might be interested in looking at some of our better universities as well—your children may wish to study in the United States at some stage."

"That would be nice," he says. "Now, we should arrange for you to return to your hotel, and I have a meeting with the minister early in the morning."

Box 9-2

Bribery and the Law

A bribe can be defined as "remuneration for the performance of an act that's inconsistent with the work contract or the nature of the work that one has been hired to perform" (Shaw 1991). Chris may well be considered to have offered a bribe to Girish with the intention of securing the contract, responding enthusiastically to Girish's hint. Maybe Pines would have won the contract anyway, but we'll never know.

Let's assume that Chris's behavior does count as offering a bribe, thus acting illegally under U.S. law. Congress passed the Foreign Corrupt Practices Act (FCPA) in 1977, in response to the disclosure of large-scale bribery of foreign officials by U.S. companies. Shaw gives numerous examples. The best-known offender, Lockheed Aircraft Corporation paid $22 million to politicians and officials, notably in Japan and the Netherlands to secure orders for its aircraft, but Exxon admitted to paying as much as $59 million to Italian politicians. United Brands and Gulf are among the other companies who paid million-dollar-plus bribes. The FCPA makes it a crime for both managers and corporations to make or offer to make payments to foreign officials or politicians for the purpose of gaining or retaining business; penalties including jail terms and fines of up to $5 million may be imposed on violators. The FCPA does not prohibit payments to government employees in order to ensure that they do their jobs, for instance, to issue permits or provide customs clearance to which the company is in fact entitled. This may seem inconsistent, but the purpose appears to be to recognize that this variety of corruption is normal in certain countries, a normal cost of doing business, and that companies who were forbidden to make such payments would be unfairly disadvantaged compared with competitors who were not. A bribe to secure that a tender, say, is treated more favorably than a competitor's is different, however, because it is an attempt to secure an unfair advantage.

By bribing Girish to give a favorable recommendation to Pines' bid, Chris violated the FCPA. Had Girish been paid merely to facilitate the normal processes leading to a fair consideration of Pines' bid, no violation would have occurred.

Critics of the FCPA charge that it hurts American companies because it puts them at a competitive disadvantage with companies that operate under less restrictive legislation. According to former U.S. Secretary of Commerce Frank A. Weil, "The questionable payments problem may turn out to be one of the most serious impediments to doing business in the rest of the world" (quoted in Pastin and Hooker 1980). They note that "In a large number of countries, payments to foreign officials are virtually required for doing business." Opponents of the FCPA also argue that the U.S. government has no business telling companies how they should operate overseas. Others take the view that different countries have different cultures and it is up to their government to determine the business environment in those countries.

(continued)

> **Box 9-2** *(continued)*
>
>
>
> Supporters of the FCPA, such as William Shaw (1991), point out that there is "no compelling evidence" that companies have lost business as a result of compliance with it, and that (as in the Lockheed case) corrupt practices are often aimed at gaining an advantage over other American rather than foreign competitors. Shaw also notes that even though bribery is more widespread in some countries than it is in the United States, it doesn't follow that it is *acceptable* in those countries. Drug dealing, he points out, is widespread in the United States but is certainly not acceptable. Norman Bowie (1993) points out that bribery of public officials "is prohibited by the laws of practically every nation" (p. 794) and states: "The notion that bribery is generally permitted and practiced abroad does not stand up to empirical scrutiny" (p. 795). There is also a concern that corrupt practices in foreign countries may spread to the domestic business environment—recall Bok's arguments about the effect of lying on honesty and trust generally. Bribery is also inefficient in the same way that monopolies are inefficient: "The bribing firm can impose higher prices, engage in waste, and neglect quality and cost controls since the monopoly secured by the bribe will secure a sizeable profit without need of making the price or quality of its products competitive with those of other sellers" (Velazquez 1992).

Sunday, September 21

After a long and tiring flight back, you take a few days off, and are enjoying a relaxing Sunday, interrupted by the phone.

"Sorry to disturb your weekend," Joe says. "Thought you'd like to know I just had a call from that Girish guy in Genala. Unofficial, of course, but it seems we've got the contract for the resort."

"That's wonderful!"

"Yes, and congratulations, Chris. One thing did puzzle me though. He said he and his wife were looking forward to meeting me and Brenda, and their teenage sons are hoping to take a look at the university. Know anything about this?"

Omigod, you think to yourself. To Joe, you say, "Hey, that's all great news. Look, I have to go, we're due at Alex's mother's in half an hour. I'll talk to you tomorrow."

Monday, September 22

When you get into work, Rosemarie has a message from Joe: You're to go straight to his office.

"What in hell is this about?" he asks. "You invite Girish to visit?"

You explain the situation. Joe is not at all happy. "Look, I went along with the car, even though it's against policy. But this is exactly what I said the policy is supposed to prevent!"

"You wanted to win that contract, you even told me to bid low. Anyway, 'When in Rome' . . ."

"Yes, but I wanted to win it on merit, not by bribery—this isn't Lockheed! Not sure if what you did is even legal. Checked out the ASCE Code—it doesn't accept 'When in Rome.'"

Box 9-3

When in Rome

Engineers in the United States argued for many years that they were cut out of overseas business because the ASCE Code of Ethics flatly forbade the use of gratuities (bribes) as a means of getting work done. In response to the difficulty of applying the ASCE Code of Ethics for overseas work, in 1963 the society revised its Code of Ethics, adding a footnote that became known as the "when in Rome" clause. It read:

"On foreign engineering work, for which only United States engineering firms are to be considered, a member shall order his practice in accordance with the ASCE Code of Ethics. On other engineering works in a foreign country he may adapt his conduct according to the professional standards and customs of that country, but shall adhere as closely as practicable to the principles of this Code."

This footnote was eventually repealed, amid great debate over the very essence of what ethical conduct is supposed to be and the relativity of ethics. If an act (say bribery) is unethical in the United States, should it also be unethical elsewhere? Or is ethical conduct relative and culturally based?

Some years ago an engineer, call him Ethan, was working for a large, unnamed construction firm in one of the Pacific Rim countries. In order to obtain a shipment of steel, he was told that the customs have to be handled by a local lawyer. Since it seemed a simple transaction, Ethan decided to do it himself. His local engineers told him that this would not work, and that he had to have a lawyer. Recognizing this as extortion, Ethan got his ethical dander up and refused to hire the lawyer. Not only did the shipment of steel get delayed, but the entire construction project was in jeopardy of failing because no materials of any kind would be released by the customs agents. What Ethan did not understand was the extensive system of paybacks and established bribes that allows material to travel from on board ship to the construction site. He subverted the system, and the locals made him pay for his actions. Eventually, Ethan was relieved of his duties by his firm and brought back to the United States.

Universalist Aarne would argue:

It the engineer is devoted to the betterment of the public good, bribery is not acceptable behavior because it allows public money to be used as payoffs that are hidden from the public. The argument that bribery is morally acceptable if everyone does it is patently false. If the majority of people in society agree that a certain ethnic groups should be slaves, this would not be morally acceptable. Similarly, if everyone in a society took bribes, it would not make bribery morally right.

Ethan, in our story, should stick to his guns and try to find a way to get the material to the job site without succumbing to bribery. If this cannot be done, then his company is perhaps doing the wrong job in the wrong country and they should leave.

Pluralist Alastair counters:

Bribery is generally regarded as wrong, at least in Western countries, because it is seen as being against the public interest, and Ethan shares this view. If the engineer is devoted to the betterment of the public good, bribery is not acceptable behavior because it allows public money to be used as payoffs that are hidden from the public. But should Ethan, in the given situation, have accepted the system as it was, recognizing that the greater good would be served?

The answer is, of course, that even if we accept that bribery is morally wrong, and this was certainly Ethan's view, bribery was necessary if anything was going to get done. So he should go along with the local culture, since his job is to get the building up, not to convert the locals to his values. He can go on maintaining that his moral values are universal, but pragmatically, if reluctantly, accept that in business you sometimes have to compromise.

We introduce the argument on the social and cultural plurality of ethics in Box 2-6, discuss it in detail in Box 9-2, and return to it in Box 14-3.

"I didn't offer Girish any bribe. I didn't offer *anything*. I just suggested he might like to visit."

"Oh, come on! What's he supposed to think?"

You're not convinced that you actually broke the law, but this is not a good time to argue about legal niceties with Joe.

"Look," you say. "You're the boss. Obviously I made what you believe is a bad decision, and I apologize for that. But it was clear what Girish wanted, and I had to make the call right there."

"No you didn't. You could have made an excuse to leave for a few minutes and called me."

At 4 A.M. your time, you think, and anyhow your experience with the Genalan telephone system suggests that this maneuver would have had limited success.

"Well, you pay me to make decisions, and I made what I thought was the right one for Pines. We've got the contract, we'll do a great job, and we'll probably get lots more work out of it."

Joe looks hard at you. "Pines is not going to pay bribes to *anyone*, never mind the situation!"

"If it's the money," you say impulsively, "I'll pay for it myself."

"For God's sake, Chris, of course it's not the money. It's the principle."

"For you, yes. But what's wrong with me inviting my friend from Genala and his family for a visit?"

"It has *zip* to do with friendship, Chris, as you well know. You bring Girish out here, you'll be doing it as an employee of Pines—or rather, as an ex-employee of Pines! Look, I'm prepared to give you a break on this, you've done a lot for the company, and I admire you for the stance you took on the Asmara even though I think you were wrong. And I also accept that you were thinking of Pines when you made Girish that offer—there's nothing in it for you, certainly. But I don't like it. So take your choice. I'll give you until 9 tomorrow to decide."

You go back to your office and think about your options. You can probably afford to pay off Girish, even if you have to sell the car. You likely won't have a job with Pines, but you'll have no problem getting another job, and Girish will probably send more work your way. You're not sure how these things work in Genala, but you figure Girish might feel that some sort of rapport has been established. This thought pulls you up short. Don't you have obligations to Girish? For all you know, Genalan officials gain most of their income in this way. If you hadn't offered Girish the trip, he probably could have got something from the competition. He's probably told his family about the trip and made plans. He's probably told his friends and business associates too—if you don't come through he'll lose face, and it might damage his career prospects. What a dilemma!

Box 9-4

Ethical Dilemmas II

We discuss dilemmas earlier (see Box 5-3). Problems that appear to be dilemmas are not necessarily insoluble. They can be resolved in several ways; here are some possibilities.

First, the decision maker may decide that one of the two principles involved is more important in the circumstances. For example, you ought to keep promises, and you ought to help people who are in need in an emergency, but what if you can't do both? Suppose you've promised to take your frail, elderly grandmother to the opera for her birthday. But as you're on your way to pick her up, you come across two small children wandering along the road—you only just manage to stop in time to avoid hitting them. You haven't seen a house for miles, so you ask them if they're lost. They tell you in scared voices that they've run away from home. They're dirty and hungry looking, and you get the impression that they've been on the road for days. No one else is around to help—your route is through a state park and it's winter. Most people would agree that you should take the children to the nearest safe place, such as a gas station, store, or house, on your way to your grandmother's house, but the children are too scared to get into your

(continued)

Box 9-4 (continued)

car. You forgot to charge your mobile phone, so all you can do is find a pay phone, call the police (and grandma), return to the children, and stay with them until help arrives. Maybe this will take so long that you'll be too late to take Grandma to the opera. You're probably going to break your promise, but you would presumably be right to do so, if that's the price of perhaps saving innocent lives. In this situation, there's no dilemma.

Second, there might be another alternative, so it isn't a dilemma after all. Perhaps, before reading on, you might want to consider what alternatives are open to you in the situation of the psychologist discussed in Box 5-3.

Sometimes, the law provides a solution, as in the famous *Tarasoff* case. Tatiana Tarasoff, a student at the University of California, Berkeley, was murdered by a fellow student, Prosenjit Poddar, who had earlier confided his intention to kill her to a psychologist at UC's Cowell Memorial Hospital. At the psychologist's request, the campus police detained Poddar briefly but released him because he appeared to be "rational." No attempts were made to warn Tatiana or her parents, who after her death sued the University for what they claimed was a breach of a reasonable duty of care. The case proceeded to the California Supreme Court, where in a split decision the court found for the parents. The majority opinion stated: "We recognize the public interest in supporting effective treatment of mental illness and in protecting the rights of patients to privacy . . . and the consequent public importance of safeguarding the character of psychotherapeutic communications. Against this interest, however, we must weigh the public interest in safety from violent assault. . . . The public policy favoring protection of the confidentiality of patient-psychotherapist communications must yield to the extent to which disclosure is essential to avert danger to others." The Court also noted that the American Medical Association Principles of Medical Ethics allows a physician to reveal confidential communications with patients if "it became necessary in order to protect the welfare of the individual or of the community" (*California Reporter* 1976).

In a well-known case in New Zealand, a general practitioner was treating a man who worked as a school bus driver for a heart condition. The doctor believed (correctly, it later emerged) that it was not safe for the man to drive, so he advised him to quit his job. The patient refused. The doctor decided that his duty to the schoolchildren outweighed his obligation of professional confidentiality, and he informed the media. The angry patient complained to the school authorities, and the doctor lost his license to practice medicine. However, there was a third option that the doctor should have taken, of informing the traffic department, in confidence, of his concerns; they would then have required the driver to submit to an independent medical examination. If he was found to be unfit to drive, he would then lose his bus driver's license (which is what happened anyway). This course of action is endorsed by the New Zealand Department of Education because it protects the traveling public without breaching the patient's privacy. Similar provisions exist in the New Zealand Psychological Society's Code of Ethics, which permits (or, in some circumstances—requires) a psychologist to breach client confidentiality in certain circumstances—for instance, when a client is believed to be a danger to self or others and no acceptable alternative to disclosure is available. The Code also requires psychologists to inform potential clients of the rules on confidentiality before the first consultation begins.

In these dilemma-like situations, the ethical decision maker may also be able to take steps to mitigate the harm. In the story involving the lost children, you could call your grandmother and quickly explain the situation, promising to arrange a later visit to the opera (whereas if you decided to ignore the children, and they became seriously ill or died, there would be nothing you could do to mitigate the harm).

How does this help the psychologist who is counseling the child abuser? As the minority opinion in the *Tarasoff* case noted, the need for professional-client confidentiality goes beyond the immediate effects of a breach: If abusers believe that they will be reported to the police or welfare agencies, they will not seek professional help and will continue to hurt their children.

One way to avoid the dilemma is for the psychologist to propose a third alternative to the client. For instance, the client could arrange to provide alternative care for his children with family members, and live apart from them until he has learned to control his violent impulses. If the client genuinely wants to change, he will probably accept this option, thus protecting the children while maintaining confidentiality and the client's trust. Of course, if he didn't want to change his behavior (and pedophiles often claim that sex between adults and children is morally acceptable), then it won't be a reasonable solution from his perspective.

Tuesday, September 23

"So, Chris, what's your decision?" Joe asks. "You in or out?"

"Just give me a moment to state my position, please?"

Joe nods.

"I admit I made a mistake," you say. "And I'm not proud of it. But that's the position I've put us in, and we have to go forward. I just did what's normal business practice in Genala—and we got the contract! Girish will be depending on me to keep my side of the bargain—I owe it to him. If I pay personally for his visit, it won't cost Pines anything. On the other hand, if we don't come through, I can see that Girish could make life quite difficult for us, and we might not get any more Genalan business. Look, Joe, that's the way they do business over there. Let's go along with it."

"No," Joe says firmly. "Don't accept this cultural relativism stuff."

"That's not the same. Those cases are about basic human rights."

Box 9-5

Human Rights I

Human rights is a controversial issue, especially in terms of the relations between developed countries, including former colonial European powers and the United States, and developing nations, especially China and Southeast Asian nations. The question of human rights revolves around the values and whether these are universal.

Are there universally true values? Or is ethics relative to culture? Engineer Aarne and ethicist Alastair have been arguing amiably about this for years and years. Here's an account of the sort of discussions they have:

Aarne: Al, it's one thing to say that different cultures have different customs, and I go along with that, but how can you say that there are *no* universal values? Ethics can't *just* be a matter of cultural relativism. Isn't there *anything* that you think is right—or wrong—regardless of what is practiced in a given culture?

Alastair: Well, I call it *cultural pluralism*. Different cultures develop different ways of living together, and by and large they work out. Think of different family structures, for instance. It's not like structural engineering, where there's just one right answer.

Aarne: Agreed, but of course you agree that the Holocaust was wrong—really wrong?

Alastair: How could I not, I lost half my relatives in it, but that wasn't the outcome of a cultural agreement about a way of life that works. What I have in mind is different value structures that promote human flourishing: There are lots of value structures, ways of life if you like, that can do this. The Nazi regime lasted for about 15 years and was at war for nearly half of that. I'm talking about different value systems that actually work, not a temporary regime.

Aarne: What about slavery then? Surely you can't possibly say that could be "right for slave-owning societies"?

Alastair: Like you, Aarne, I'm a product of my culture, and of course I disapprove of slavery. But again, it was a temporary phenomenon in America, and I think it was inherently unstable. Also, it took a century before civil rights legislation gave African Americans anything resembling equal rights—longer than the period of slavery after independence. And in ancient Rome, unlike the Old South, almost every citizen had a slave, and the alternative for most of those slaves was starvation.

Aarne: Your position works for different customs, and I can even see how a humane system of slavery could be better than the alternative in societies with economies very different from our own. But this is about right and wrong. How do you feel about genital mutilation of girls in Africa and child prostitution in Thailand?

Alastair: So far as child prostitution is concerned, of course I disapprove of it, and so do Thai people—

(continued)

> **Box 9-5** *(continued)*
>
>
>
> it's foreigners that spawned the trade. But again, I'd say that it's the product of a particular historical situation that creates economically dispossessed people—a situation, incidentally, that was produced by the same Western nations that bang on pompously about human rights to the very nations that they have impoverished. According to their own ideals, Western countries should be doing something so poor families don't have to sell their children into prostitution instead of preaching to them about human rights.
>
> **Aarne:** Just because some Americans and Europeans are hypocrites doesn't mean child prostitution is OK. But you didn't say anything about genital mutilation . . . and that isn't just a temporary phenomenon in a society that is going to collapse like Nazi Germany did.
>
> **Alastair:** You've got me there. Yes, I think it just is wrong and I can't see any possible justification for it . . . but it's tied up with entrenched belief systems, and successful programs to get rid of it have been based on education at village level, not international demands by Western countries that it be stamped out. All cultures can change, but they have to do it their way.
>
> **Aarne:** Oh, so it's OK for genital mutilation to continue for a couple of generations more of girls in Mali and Somalia to undergo it while we wait for the process of internal cultural change?

"No, Chris, ethics isn't just about basic human rights—anyhow, it goes against my principles to pay bribes, whatever other people might think. And what's more, I don't buy into that stuff about me not being responsible. Last night I went to Blockbuster and rented that old movie *Judgement at Nuremberg* . . ."

"Great cast! Spencer Tracy, Burt Lancaster, Dietrich, Judy Garland, Montgomery Clift," you automatically butt in.

"Yes, yes, and Richard Widmark and I bet you didn't know this: William Shatner."

You didn't, but you are not about to admit it. He continues:

"And that convinced me that if I condone this, I'm responsible for it. As for the problems you mention, the difficulties in getting the resort built, it'll be up to you to deal with them. Personally, I don't believe there'll be too many problems. The Genalan government wants this project completed successfully, and anyhow if the system's as corrupt as you apparently think it is, Girish might be out of favor next week."

"So, what's it to be?"

You know Joe well, and you have already decided what you are going to do. "Thanks for hearing me out, Joe. I'll explain the situation to Girish and hope he understands."

"Just what I wanted to hear! I'd hate to lose you, Chris. Now, go do such a great job that we'll be setting up a Genalan subsidiary next year. How does that sound?"

With me in charge, you think. But what are you going to say to Girish? You can hardly sell the car to pay for his trip—Joe knows how much you and Alex love it and he'll certainly wonder.

Friday, October 17

You consider the financial side of your problem, and decide that there is a way to buy out of your difficulties. A few years ago you bought some stock in a dot-com start-up, at $2 a share, and recently sold most of it at $64. It's almost "mad money" to you, and it will nicely cover the plane tickets for Girish and his family. You figure it's a good financial investment.

If the job comes through and Pines makes a handsome profit, Joe will undoubtedly share this with you as a Christmas bonus, so you figure you'll get it back. Besides, how can this be thought of as a bribe? You are just helping a friend finance his trip to the United States. Nothing unethical about that. But if it is all above board, why, you wonder, would you not want to tell anyone about what you are doing? And most importantly, you still have to figure out what to tell Joe and Girish.

Saturday, October 18

Alex and the kids are visiting Grandma again, and you're suddenly reminded about the nice day you had at the coast around this time last year. You decide that you'll go there again. Maybe a plate of BJ's crabcakes and a good long walk on the beach will clear your head and you'll be able to make some plans.

The trip starts off inauspiciously: BJ's has closed and is all boarded up. Nearby, there's a new place, Kim's Krabkakes, which has a large chalkboard outside promising "The Koast's Keenest Krabkakes!!!" and "The Best Widescreen World Series Coverage on the Koast!!!" You decide to book into a motel and sample the "Finest Krabcakes." The crabcakes are pretty nice, but they're not up to BJ's, who, the bartender tells you, met someone with a yacht and was last seen heading for Barbados.

After lunch, you walk along the beach, watching the pelicans and waiting for ideas. After a while, you sit down, and pull out your notebook.

You decide to make up two stories.

In a few weeks, you'll tell Joe that Girish has acquired some funds to pay for his trip. You'll be a little vague about the source, and he'll understand that, since you can't be expected to be an expert on the ins and outs of Genalan business and politics. You'll even hint that you suspect that there may be something not quite kosher about Girish's source of funding, and leave Joe to draw whatever conclusions he wishes. He might think, say, that Girish has extorted kickbacks from contractors, but he can hardly object since it isn't within Pines' control.

Your second story is for Girish, to whom you'll explain Joe's scruples, adding that "I've found a way to fix it." Girish won't be interested in *how* the visit is being paid for, so long as it *is*.

Pleased with your ingenuity, you retrace your steps, take a shower in the C-Breeze, and stroll down to Kim's to watch the Yankees get lucky for the fourth time.

Discussion Questions

9-1. As we present it (see Box 9-4), Chris's situation appears to be a true dilemma. Are there further options for Chris?

9-2. Under what circumstances would the engineer not be bound by confidentiality? Devise a case study in which the engineer would be required to breach client confidentiality in order to act morally.

9-3. Consider this argument for competitive bidding: In a market situation, we expect to shop around for goods and services, and price is certainly a consideration for most of us. But obviously we don't always buy the cheapest product; otherwise, everyone would be wearing shoes from Wal-Mart. What we seek is value for money. Why should engineering services be marketed any differently than footwear?

9-4. The FCPA distinguishes between payments made to government employees to provide legitimate services and payment to improve the prospects of winning a contract. Ethically, is there any difference between these two types of payments?

9-5. If Chris used private funds to finance Girish's trip, is this unethical? Why does Chris not want to tell others the truth? What is troublesome (ethically) about this arrangement?

References

Bowie, N. 1993. "Business Ethics and Cultural Relativism." In *Business Ethics*, T. I. White (ed.) p. 790–799.

Pastin, M. and M. Hooker. 1995. "Ethics and the Foreign Corrupt Practices Act" in *Business Ethics*, W. M. Hoffman and R. E. Frederick (eds.) New York: McGraw Hill, p. 539.

Shaw, W. H. 1991. *Business Ethics*. Belmont, CA: Wadsworth, p. 265.

Tarasoff v. Regents of the University of California, California Supreme Court, 1 July 1976, 131. *California Reporter*, pp. 14–33; dissenting opinion, pp. 34–42.

Velazquez, M., SJ. 1992. *Business Ethics: Concepts and Cases*, 3rd ed. Englewood Cliffs, NJ: Prentice Hall, p. 196.

10

Professional development of others

Friday, November 15

Joe accepts your story, and he is happy for Girish to visit Pines again. As you predicted, Girish doesn't ask any questions.

Over the weekend, you spend an afternoon preparing for your presentation on engineering excellence. It seems a long time since last October, when you were stuck in a traffic jam and thinking about engineering excellence. Looking back over your old notes, you wonder if you could honestly say all the things you intended to say, in view of subsequent events, and you ask Alex's opinion.

"Figure it this way," Alex said. "The point is to present students with ideals to aim at, don't focus on the negative. That's what we try to do at the business school, at least until the goddam president closes us down! Sure, everyone has to make some compromises for their career—and their employer. But there's plenty of time for the students to find that out when they start work. You don't want them showing up on their first day in their first job all cynical about engineering professionalism!"

You're uncharacteristically nervous as you drive to the university for your presentation, but it all works out well. You begin by briefly telling the students a little about yourself and your career. You describe some of your projects and let them ask you about what you do on a daily basis. They are interested in what aspects of their education are useful in real-life engineering.

Toward the end of your presentation, a student says, "You obviously enjoy your job—what exactly is it that you really like about engineering?"

Box 10-1

The Existential Pleasures of Engineering I

Like all books, the book you are now reading is partly autobiographical. I (ethicist Alastair speaking) had a humanities education and knew little about science and technology until I met my coauthor (engineer Aarne). I came to understand the existentialist pleasures of engineering when I was visiting with Aarne. His (then 7-year-old) daughter was away at summer camp, and I was using her room. The room had a rainbow painted all around the walls and I was very impressed that the rainbow looked right from every angle, especially how it fitted into the corners of the room (you try it!). Aarne said that it was very hard to do, he knew he'd solve it, and he felt great when he got it to look just right. This is probably why Joe loves engineers.

You expected a question along these lines, and as you planned, you tell the class that for you, engineering provides much more than a good career. You mention Samuel Florman's lyrical writings on the great intellectual and spiritual rewards of engineering. But, you note, Florman passionately argues that engineers have no special ethical insights, and therefore they have no ethical responsibilities beyond providing excellent client service. In particular, he says, it is not the engineer's job to pass judgment on the social aspects of engineering projects.

It's up to you, you tell the class, to decide whether you agree or disagree with Florman—make up your own minds!

The students seem interested and appreciative, and you feel good about your presentation.

Box 10-2

The Existential Pleasures of Engineering II

In his book *The Existential Pleasure of Engineering* (1972), Samuel Florman argues against the view that engineers have special social responsibilities. Engineers are experts on what they are trained to do—to design and build structures, machines, systems—and they can benefit humanity by making their expertise available to those who need it. As citizens, engineers have their own values and understanding of society's needs, but that is all: As engineers, they are not experts on society's needs and should not presume to tell people what is best for them. Recognizing this frees engineers to enjoy the great personal—spiritual even—rewards of engineering: the joy of engagement with technical problems, and producing simple and elegant solutions to those problems. Engineers also improve people's lives—not by providing them with what engineers think is good for them, but by helping them to achieve their goals.

Florman's view of engineering ethics may be one of the reasons engineers are so often the guys in the black hats, like the bad gjuys in cowboy movies. Consider this example: Engineer Kurt Prufer used his engineering skill and common sense to build the ovens at Auschwitz, eventually expanding them and the associated ancillary processes to be able to gas and incinerate over 200 people a day (Vesilind 1994).

Prufer did a technically excellent job, and certainly he did nothing illegal. He simply followed the dictates of his superiors, his clients, and society. He trusted "others to assess those consequences . . . beyond the engineering task" (de Rubertis, 1994). If nothing else, Prufer was politically correct by the standards of his society.

De Rubertis quotes Florman's example of two musicians discussing whether or not they should play music for a political purpose or for the enjoyment of others, implying that engineering should be thought of in the same manner. The flaw in Florman's argument, it could be said, is that music and engineering are fundamentally different. Seldom if ever has music caused any harm to anyone, and a decision to play or not to play has had few if any serious consequences. Still, like other works of art, music is not politically neutral. Like other artists, composers and performers often seek to use their art to make political points. Consider the values promoted in popular music, for example, by the Spice Girls (the idea of "girl power"), in the Beach Boys' celebration of the 1960s beach culture ("California Girls" and "Back in the USA"), and in more recent versions of youth culture—Guns 'N' Roses ("Paradise City") and Beastie Boys ("Fight for Your Right [To Party]"). In classical music, Benjamin Britten's *War Requiem*, was a purely political (pacifist) work while Tchaikovsky's "1812 Overture" is a celebration of a battle won. Even if the artist has no overt political agenda and he or she is only carrying out a commission or responding to a current trend, music may have a value, even a political content. In Israel, Wagner is certainly not regarded as politically neutral, and for many years his works were not performed because of a presumed association with Nazi Germany.

(continued)

> **Box 10-2** *(continued)*
>
>
>
> Accepting all this, it is very clear that engineers, by virtue of their training and skill, can and often do cause much more direct and irreparable harm to people, cultures, and the environment. Nothing gets built without engineers. Engineers are an indispensable component of a civilized society, so it can be argued that they have the most clear and compelling responsibility to do "the right thing"—which may not be what the client wants.
>
> Engineers who take this view might say: Yes, if all of us *did* decide to undertake only engineering projects that we felt were the right thing to do, then some things would *not* get built. What would have happened if Adolf Hitler could not have found a single engineer to design and build the ovens at Auschwitz?

You have a special fondness for and a keen interest in De Tocqueville University (DTU). As a graduate of the School of Engineering only 12 years ago, you were flattered when you were asked to become a member of the Dean's Council, a group of industrial supporters who meet twice a year to review the academic programs and to offer advice. Everyone knows, of course, that the primary reason for the Dean's Council is to assist in fundraising, and indeed the members of the Council have been active in finding sizable gifts for the School of Engineering. So it came as no surprise when you received an e-mail from the dean:

I understand you are making a presentation to one of the engineering courses on Thursday. How about you and I have lunch afterwards?

Ivan
****
De Tocqueville University
School of Engineering
groznyi@dtu.edu

You are glad he wrote, because this will be an opportunity for you to find out more about how to get Girish's kids into DTU.

Ivan has chosen Al's Diner for lunch. It's a fiendishly expensive supposed re-creation of a late '50s–early '60s restaurant, on a grand scale. The waitpersons, including the bobby-soxer who shows you to your table, roller-skate along the polished wooden floor with orders of burgers, fries, Cokes, beers. The booths are fiberglass replicas of Thunderbirds, Galaxies, Impalas. You're a few minutes early and you wander over to the replica chrome-and-neon Wurlitzer. The Fleetwoods' "Mr. Blue" is playing. On a small table next to the jukebox is a bowl of nickels. You feed in three, push buttons at random, and return to your table where you discover a bottle of Dr Pepper. Leaning against it is a card reading: "Gosh, thanks for visiting Al's! Here's a little gift to start your meal off right!"

Ivan walks in as Guy Mitchell begins to sing "She Wears Red Feathers and a Hula Hula Skirt"—surely you couldn't have selected that, not even randomly?

He greets you with, "Hey, how about those Giants?" You remember that Ivan used to teach at San Francisco State.

"Best team money can buy. Now Atlanta, they build their team, they don't just go out and buy them."

"Oh, right, so how much they paying Glavine and Maddux? Maybe he looks like an accountant but he sure don't need one to get rich. Money makes the world go round—baseball, education, you name it."

You do some more small talk about sports, family, and mutual acquaintants. Ivan is giving his full attention to his 16-ounce New York cut with fries and onion rings, which rather dwarfs your Reuben.

"Hey," he says. "This place reminds me of the thirty-fifth anniversary celebration at DTU. It's a legend, you hear about that ever?"

"No, tell me about it."

"What happened, hundreds of alumni showed up, and their families, it was a real big show. This is a real optimistic time, right, defeat the Commies, make the world safe for democracy, et cetera. Also, obtain lots of alumni money for DTU. Seems they laid on a real nice dinner and reception and a dance, Patti Page was the headline. But the president thought he was pretty cool, so he asked the students to hire a band too so the parents could see what a good relationship there was between students and faculty. But the student president had a real grudge against the whole DTU hierarchy, and he got a really raunchy band that just grossed everyone out. Musta been a great laugh, but I doubt it did much for the endowment."

"I didn't realize DTU was that recent—must be, what, only 60 years old? How did it get started?"

"Hey, you could say we're the beneficiaries of the casualties of war. See, this Theodore Russell II, a mechanical engineer he was, from RPI, made a huge fortune from munitions early this century, but then there was the First World War and he was just sickened by the slaughter. At first Russell sold stuff to just anyone, then he sees guys on *both* sides dying from *his* munitions. In 1916, he suddenly sells his business—huge profit, I should add.

Box 10-3

Engineers and Armaments

I (engineer Aarne speaking here) knew an electrical engineer who used to be employed by a large corporation and had resigned to open his own computer consulting practice. He took a large cut in pay and job security by resigning from what appeared to be a lucrative and challenging engineering job. One day I asked him why he did it.

"I found out that I had been misled," he explained. "I thought the project we were working on was to develop a commercial laser application for consumer products. Instead, it turned out that I was working on a project for the Department of Defense that could turn out to be a new and more fearsome weapon of

(continued)

Box 10-3 (continued)

mass destruction. I did not want to be a part of that and resigned."

I could not get any more specifics from him, and he did not want to explain further. He had been working with security clearance, and he felt that he should keep that promise to not talk about his work.

This problem is widespread in industries that hire engineers and put them to work on small projects that have no obvious connection to anything else. Often it is impossible, for various security reasons, to find out what the large project is. One only knows about a small piece. Similarly, the use of a building under construction might be unclear. What looks like an ordinary office building may eventually be used for public executions, for example.

Is it ethical for engineers to work on things that are designed to kill people? Of course, most of the buildings and gadgets engineers design can and often do kill people. But we are talking here about things that are *designed* to do that—things like bombs, guns, land mines, poisonous chemicals for warfare, and so on. None of the codes of ethics are explicit about working on projects that are designed to kill people. The closest we come is the now familiar first canon:

"The engineer shall hold paramount the health, safety, and welfare of the public."

The "public" in that statement might mean the public for whom the engineer is responsible. Consider an engineer who conceived and designed the dum-dum bullet, the so-called "cop killer" because it can penetrate protective vests. The development of this bullet was a response to market forces. That is, if it was manufactured, people would buy it. It had no conceivable use other than to kill people wearing "bulletproof" vests, which is almost always law enforcement officers (although they could be terrorists as well). The "public" was the people who thought they needed protection from the cops and were ready to kill them if necessary. Can we argue that the engineers who worked on this project were acting immorally?

Of course, one could argue that in the case of the ammunition that penetrates Kevlar ("bullet proof") vests, it is not the technology that is immoral, but rather the use of the technology. Conceivably, such ammunition could have legitimate uses. Even technology designed to kill people can have a legitimate use

and it is the control of technology that is important, not just the technology itself. The problem is that in the United States, Russia, and many poorer countries, legitimate military technology such as semiautomatic rifles are available to maniacs and street scum. In Cape Town, South Africa, one can illegally but easily purchase Czech AK-47 assault rifles for less than U.S.$10, with a box of 7.62 ammo thrown in. This is because the Soviets flooded Angola and Mozambique with these firearms during their civil wars. Mikhail T. Kalashnikov, the inventor of the AK-47, probably did not ever dream that his rifles would end up being used by rival drug gangs to kill innocent citizens and each other.

Consider now the manufacture of military aircraft such as the B-2 bomber, which has no conceivable civilian use. These aircraft are designed to kill a different "public"—whomever the owners and operators of the aircraft deem necessary to kill. If the aircraft belongs to the people of the United States, for example, then they are expected to kill non-Americans. Would the destruction of these people be morally acceptable if their death enhances the health, safety, and welfare of Americans?

The so-called Gulf War, fought in response to Iraq's occupation of Kuwait, did not endanger the health or safety of Americans, but welfare was certainly important. We in America have grown used to cheap petroleum products, and the issue was whether Saddam Hussein would be allowed to have control over a substantial portion of the world's oil reserves. The B-2 and other offensive weapons played a major role in the destruction of tens of thousands of Iraqis.

But suppose the word *public* in the code of ethics really means all humanity. What then is the engineer's responsibility, and would it preclude working on armaments that are designed to kill some fraction of the public?

I (engineer Aarne speaking here) once met another engineer who worked for Sandia Labs, the primary research facility in the United States for armament development. He was an older engineer who had had a successful career. His greatest achievement, it turned out, was the development of a sonic triggering device for bombs. Older-style bombs had a piston that

(continued)

> **Box 10-3** *(continued)*
>
>
>
> when it came in contact with a solid surface, forced the rod in and caused the bomb to explode. Sometimes the piston stuck, or was bent, and the bomb never exploded. The sonic device had no moving parts, and the vibrations of the bomb hitting something were adequate to cause it to detonate. He explained this with such enthusiasm and such pride that I did not have the courage (not to say manners) to ask him how he felt about an entire career in engineering spent on finding better ways to kill people.

"Then he meets this Quaker who converts him to the cause of universal peace, and he starts campaigning for pacifism and the end of war, gets investigated by the government, and nearly goes to jail. Then he gets real unlucky and dies in the influenza epidemic.

"Story goes, on his deathbed—this is 1919, right?—his only family, Rebeccah and Elizabeth, that's his wife and daughter, are sitting round the bedside in the hospital, and he raises himself up from the sickbed and goes, 'I have a vision! A university of peace, justice, and international understanding! Please, please, find a way.' Then he dies. So there isn't time for any formal document or anything, but Rebeccah and Elizabeth decide to honor his last wishes, talk to all kinds of people—some say including Woodrow Wilson and Henry Ford—and then they set up De Tocqueville right here."

He smiles. "Hadn't been such a bloody war, Russell might have gone on happily manufacturing deathtech. So you could say, like they do on the war memorials, 'They Did Not Die In Vain.' Every cloud has a silver lining. Like I said, money makes the world go round. Nothing really changes."

Including Ivan's stock of clichés, you think.

"So how did your lecture go? Students give you a lot of trouble?"

"No, they were wonderful. It must be a great job, to do that all day . . . to teach students. But it sure does drain you. I doubt if I can do anything else productive for the rest of the day."

"Yeah, that's why some of us smart professors got into management," Ivan smiles.

"I guess deaning doesn't leave you much time for students."

"No, I really don't have much to do with students. Why do you ask?"

"Today we got to talking about the School of Engineering, and it turns out the students don't even know who you are. They've never had any contact with you or even seen you."

"Guilty as charged. The only time they're likely to is at graduation. I really don't get involved in the educational programs. I leave that to the professors and my associate dean. My job's to raise funds so they can do the research they want to do."

You want to argue that the role of the dean should be primarily and foremost to care about the educational experience of the students, but you think better of it. You've heard the dean's views on this before, and you wonder what is to become of American engineering education if all the deans have this attitude. You care about the undergraduates you met today, and wish there was something you could do to change the views of the dean. But an expression flashes into your mind: Never try to teach a pig to sing. It won't work and it annoys the pig.

"What I wanted to talk to you about," Ivan continues, "remember I told you about the lab we were going to construct, that a rich alumnus was financing? All signed and sealed, but he had this tax problem, couldn't pay us anything 'til next year."

"I thought it sounded risky at the time."

"Hey, it wasn't risky at all, we had it in writing, he was going to pay a million twelve and we went ahead and let the contract. Next thing we know, the guy shoots himself in the head and it turns out his company is in terrible shape, really it's bankrupt, no way it can trade out of it, leveraged to the hilt. Sell the plant and stock, there won't be a dime for us."

"Doesn't your agreement with him entitle you to anything?" you ask.

"No way in this state. Our bankruptcy laws, the secured creditors get priority and they'll be lucky to get 50 cents on the dollar. Us, we get nothing."

"So how are you going to finance the lab?"

"Same way we paid for the extensions to the library, I hope—corporate sponsorship." He looks at you speculatively. "With naming rights of course."

"Who paid for the library?"

"A fast-food chain. Ever hear of The Veggie Patch? It's vegetarian fast food. The founder doesn't believe in killing animals, so he has set up this chain of vegetarian restaurants—pumpkin burgers, rutabaga barbeque, God knows what."

"You're telling me your new library is called The Veggie Patch Library?"

Box 10-4

Vegetarianism

Aarne: There is nothing wrong with eating meat.

Alastair: We humans have a choice and since we are omnivorous, we can survive on nonanimal food. No need to eat meat.

Aarne: But this begs the question of why we *ought* to not eat meat. Of course, there is wasteful and conspicuous consumption—seemingly a singular attribute of humans—but unless such wasting breaks some moral rules, there is nothing wrong with it. More to the point would be the rights of nonhuman animals to not be caused to suffer unfairly, or to be deprived of their supporting environment. But if this does not occur, there is nothing immoral about eating meat.

Alastair: I believe that eating animals can simply be *wrong*. We humans are not like lions in that we do have a choice about whether or not we eat meat. We can choose to be vegetarians, unlike the lion.

Aarne: If something is claimed to be wrong, it has to be proven wrong. If eating meat is wrong, then you have to show me how, using logic, this is the wrong thing to do, much as we can show that lying is wrong.

Alastair: I don't need to prove to you that causing pain is wrong unless there is an exceptional reason, such as a surgical operation.

Aarne: I agree that the pain and suffering that a calf has to go through to become veal is horrendous, and I will not eat veal for that reason. But in general, I cannot justify not eating any meat on the basis of pain to the animal.

Alastair: Another way that we can show that eating meat is wrong is to do an energy balance. Growing grain to feed to cattle is only 10% as efficient in providing nourishment to humans as eating the grain directly, because the cattle need 90% of the energy from the grain for their metabolism. If we then agree that reducing the impact of humans on the earth is good, then it follows that the least energy use can be achieved by being a vegetarian. In addition, animals (especially sheep) produce huge quantities of methane, which is a greenhouse gas and has been implicated in global warming. There is no doubt that if the objective is to reduce the detrimental effect on the environment, we should all be vegetarians.

Aarne: But again you assume that each one of us should sacrifice our own wants (e.g., no more Big Macs) for the sake of some larger goal like energy efficiency and the survival of the human species. It seems clear that market forces will control energy availability, and we should not reduce our individual energy use until forced to do so by price increases. There would be nothing wrong with being a vegetarian if everyone did it, but it does not feel good to sacrifice some of the quality of my life when everyone else is not doing the same.

> **Box 10-5**
>
> ## Reverence for Life
>
> Albert Schweitzer (1875–1965) believed that all life has value and that we ought to develop in ourselves a "reverence of life" (1933).
>
> Schweitzer was an immensely talented person. Not only was he the best organist and interpreter of Bach's organ music in the world, but he was a skilled physician as well and an ordained minister. In the middle of what appeared to be an exemplary career, he chucked it all to go into deepest Africa and set up a hospital for the natives who up to that time had no modern medical care. He came back to Austria occasionally to play organ recitals so he could earn enough money to pay for the hospital.
>
> During one of his trips up the Congo River, Schweitzer had a revelation. As he told it: "Late on the third evening, at the very moment when, at sunset, we were making our way through a herd of hippopotamuses, there flashed upon my mind, unforeseen and unforethought, the phrase 'Reverence for Life.'" By this he meant that not only is human life sacred, but *all* of life is sacred. In Schweitzer's view, all life has a right to exist, and we humans should not prevent other creatures from achieving their full life potential.
>
> Applying Schweitzer's "reverence for life" idea to engineering results in some interesting conclusions. Much of what engineers do affects other creatures, and if engineers believe in "reverence for life," these actions should be tempered by the rights of other living things.
>
> Consider what would happen if this philosophy was included in the engineering code of ethics. The first canon of the engineering code of ethics might then read:
>
>> The engineer shall hold paramount the health, safety, and welfare of the public unless the rights of other living things are unfairly compromised.
>
> We have added the word *fair* because all living things depend on each other in some way, and it is the need for survival that is important. Thus, a human is not expected to stop eating all formerly living things, no more than a beaver is expected to stop cutting down saplings or a lion is expected to stop killing zebras. The point is, however, that humans seem to be unique in their conspicuous waste of resources. Is it respectful to formerly living things (in terms of Schweitzer's reverence for life) if we are wasteful with our food?

"Naw, he named it after Albert Schweitzer, famous humanitarian I'm told. The donor insisted on a bronze plaque in the foyer that summarizes Schweitzer's philosophy of reverence for life. Hope he never gets to see the germ warfare section in the library, haw haw."

Why do I like this man, you wonder. Well, Hannibal Lector was a charming man as well as a mass murderer.

"You ever see *Silence of the Lambs*, Ivan?"

"What? Oh yeah, I do look a little like Anthony Hopkins, my wife says. Why do you ask? You thinking of setting up a chain of cannibal restaurants?" he laughs.

"Anyhow, the provost wants that lab built and when I was wondering how, I suddenly thought of old Joe, he's a bit of an engineering groupie, yes? The Pines Structural Engineering Laboratory? Oil painting of Joe in the foyer et cetera? Joe memorial annual molto prestigious lecture when he passes on to the great beyond?"

Not that much of a groupie, you think.

"With your tuition rates and the endowment, I'd have thought you could fund it OK," you suggest.

"Hey Chris, there's nothing in the endowment for construction. All our operating funds go to pay salaries and to run the place."

Box 10-6

The Ethics of Asking and the Ethics of Giving

Philosophers and behavioral scientists have debated for many years whether or not the human being can be truly altruistic. Is it possible for people to give of their own wealth or time in such a way as to expect nothing in return? Those who believe that humans can be truly altruistic point to the many anonymous gifts given to civic and educational organizations. Surely, the giver in such cases expects nothing in return since few people would even know of the gift. But, counter the cynics, the reason the gift was given was because the giver wanted to feel proud, or self-righteous about the deed. The reason for the gift was that it made the giver feel good, and thus the gift was not altruistic.

Many organizations such as colleges and universities in the United States depend increasingly on the generous support of their alumni and corporate friends. Conservative estimates are that the tuition students pay in private colleges covers only about one-third of the actual cost of their education, and the difference comes from various charitable sources. State-supported universities also rely on giving, with the state making up the difference between the private university tuition and the much lower tuition at state-supported universities. As this dependence on contributions increases, universities are increasingly beholden to the givers. Most corporations have gift-giving foundations, and the funds expended by these organizations are used by the corporations to influence the direction of research and education as well as to demonstrate corporate generosity.

The university can "sell" the naming of buildings, departments, professional schools, or even entire universities. Small colleges that changed their names as the result of a single large gift include Duke University, Bucknell University, and most recently, Rowan University. Once a university has named a building or program, it cannot ethically turn around a few years later and name the same entity after a new donor. Many universities are "named out," with few buildings and programs left without a sponsor's name attached to them.

The ethics of giving and taking can be complex, and problems can scuttle a potential gift. For example, some years ago a large foundation wanted to give $10 million to Duke University to establish a bioengineering center, and asked only that it be allowed to approve the faculty to be appointed to the center. The gift was rejected by Duke because the administration believed that giving up the right to name its own faculty was too precious and this was not for sale. Undaunted, the foundation then gave the money to MIT, which did not seem to have the same problems with faculty appointments.

In another instance, a donor wanted to give money to build a large museum at a university and selected a site for the building. Unfortunately, the site was right in the middle of a field being used by several botany faculty as a teaching laboratory. The development personnel tried to convince the faculty to move their lab elsewhere, but they would not budge. The donor decided that if the university could not control its own faculty, it did not deserve his money.

Corporate gifts can also diminish the independence of a university or school. At Duke University, the Coca-Cola Corporation sponsors a seminar series at the School of the Environment. The corporation participates in the selection of the speakers and thus indirectly affects the agenda at the school—precisely their motive. Incidentally, the dean of the school did not find it humorous when some students showed up at the seminar wearing Pepsi-Cola T-shirts.

Nike, the makers of sports shoes and other sporting equipment, now has exclusive rights to sell such equipment to a number of universities, including the University of North Carolina. For this privilege, Nike donates $7 million annually to the athletic program. Similarly, IBM has an exclusive deal to provide computers to the same university, again contributing millions to the general fund for the privilege. The loss of independence of both the academic and athletic programs is considered minor by the administration, but the threat of losing this support must affect the decisions and activities of the university.

Some would argue that there is a legitimate cause for concern. Suppose a corporation established a chaired professorship (now commonly selling at $1,500,000 a pop) in engineering economics at a large school of engineering. Would they not expect some
(continued)

Box 10-6 (continued)

compatibility with their political and ethical outlook on engineering economics? Depending on their political learnings, would they be upset if the university hired a Marxist economist to fill the chair?

On the other hand, while corporate giving might be self-interested, it is not necessarily evil. There is much good that comes from corporate support of education, and many donors expect nothing in return, nor do they want to influence the policies at the universities. They believe that donating money to universities is simply the right thing to do.

Trying to steer the conversation away from the subject of Pines' possible involvement in the lab, you ask, "Does the school have scholarships for people who can't pay for it themselves?"

"Yeah, we have to give 10% free places—that's fees, room and board, textbooks, everything—to, I quote, 'those who due to their unfortunate financial circumstances are unable to avail themselves of the advantages of higher learning.' God knows who he was thinking of—DTU policy is, go for minorities, and each school has to meet the 10% target. We do that, no problem. The second thing is, we're supposed to have 10% of our students from developing countries, the international understanding bit, right? Rebeccah and Elizabeth built in this special thing to encourage us, for each foreign student who comes to DTU from less developed countries—there's even a *list* of them, believe it or not—we get the full cost of an "unfortunate circumstances" student paid from some special fund. That's an extra $25 K for each of those foreign students that we get."

"Sounds like a good deal."

"Yeah, but well, we don't often meet the 10% students from developing countries quota. Next year we're looking at maybe 20 short. Of course," he quickly adds, "this is all just between ourselves, right? I mean, we wouldn't want people knowing the school can't attract enough students. Problem is, rich foreigners mostly go for real prestigious engineering schools like MIT, or colleges in small towns where they feel their kids will be safe, and very bright students from those countries get scholarships to Princeton or some such. Hence, we have difficulties recruiting students from those countries who meet our admissions criteria."

There is a short silence, which you suddenly break, as if an idea has just struck you. "Well, I don't know if this is any help, but I've been doing a lot of work"—well, some work—"in Genala, and I know a number of people"—one is a number—"who are keen to send their kids to study at a good American engineering school. I take it Genala is on the list?"

"Interesting," Ivan says, and you can almost see the wheels turning as he calculates multiples of $25,000.

"I have to go back there in a couple of weeks. Send me some literature about DTU and the engineering school and I'll see what I can do to recruit you some students."

"Hey, thanks Chris!" Ivan says, reaching for the check.

CHAPTER 10

Monday, September 4

Everything seems to have worked out. Girish's twins are studying at DTU, as are three of their friends from Genala.

You are again having lunch with the dean, at his invitation. He picks you up at work and, on the way to the restaurant, tells you that your plans for the education of the twins have worked out very well. "When those guys go back to Genala, tell their friends about DTU, who knows? Girish faxed me yesterday, he's suggesting a long-term exchange scheme, could be 20 Genalan students a year. We're talking a million a year!"

Once you're seated in the restaurant, he continues to rave on about the Genalan connection that you've established: "I can't thank you enough, Chris!"

"This lunch will do for starters," you say, glancing around at the palatial restaurant. Freddie's has won numerous awards and is almost as hard to get into as DTU.

But Ivan evidently has connections too. Shortly after you arrive, a very tall man in a midnight blue tuxedo, evidently Freddie himself, strides flamboyantly to your table carrying a bottle of Bollinger and four champagne flutes and greets Ivan like an old friend. Ivan explains that he and Freddie grew up together in Passaic, New Jersey. They both dropped out of high school, worked in a variety of odd jobs, and served in the same unit in Vietnam. After their tour of duty ended, they went their separate ways: Ivan to North Carolina State and then Lehigh, where he received his engineering degree, and Freddie back to Passaic to work in a small restaurant near the railroad station. "But we always kept in touch, eh Freddie?"

"Course we did," Freddie booms, "and I owe it all to Ginge."

"So where is she today?"

"Here any minute!" And sure enough a slim, elegant, blonde woman wearing a long black dress appears almost immediately at the table. She is almost as tall as Freddie.

"The secret of my success, of course, was marrying the boss."

After more small talk, the pair leave to greet more newly arrived guests, and bearing another bottle of Bollinger. You recognize one of the guests, but you can't quite place him.

You're curious about Freddie and Ginge, and say, "Nice couple. But are those *really* their names? She looks like a natural blonde to me."

Ivan smirks. "She sure is. Her real name is Beatrice and he was Norman at school. But can you imagine those names on a restaurant?"

Ivan abruptly changes the subject. "You see that guy in the dark suit, just walked in? Recognize him?"

"Yeah, but I can't place him."

"Trey Wilson, he was Shoeless Joe Jackson in *Field of Dreams*."

"No, that was Ray Liotta. Trey Wilson was in *Bull Durham* with Susan Sarandon."

"Oh, right, and Kevin Costner was in both of them."

"True."

"Anyhow, the Girish's twins are doing real well, though their friends, er . . ."

"Are having some trouble keeping up?" you inquire.

"Yeah, now that you mention it." He lowers his voice, though the tables at Freddie's are so far apart that he'd practically have to shout to be overheard. "Just between you and me, right, we have to put some pressure on the faculty to occasionally stretch a point or two to keep them from flunking out. Doing our bit for international goodwill, I call it. Plus the extra money has been enough for us to finish the lab."

You're not sure what to say. You strongly believe that an excellent school like DTU should take in only excellent students—and you're relieved that the Girish twins got in on merit, but you are concerned that other students are kept from flunking out by pressuring faculty to give them passing grades. What kind of engineers would they be, and who worries about the quality of engineering education?

Ivan must have read your thoughts, because he continues, "Anyhow, our esteemed president's really into this affirmative action stuff. Like, as well as the foreigners and the poor, he wants more women students and faculty. He's even thinking about a quota system and a hiring freeze on male faculty. Ask me, it sucks, but there you go."

Box 10-7

Maintaining the Quality of Engineering Education

One of the hallmarks of a profession is that it itself controls entrance to the profession. Physicians, for example, run the medical schools where young doctors are trained. Physicians also organize examinations for medical school graduates, and they themselves design the exams and grade them. It is physicians who decide who become new physicians.

Engineers in the United States are trained in schools that are accredited by professional engineering associations and an umbrella engineering organization called Accreditation Board for Engineering and Technology (ABET). Although this organization arranges for the visits to engineering departments and oversees the accreditation procedure, many of the standards for each individual engineering discipline are set by the society for that discipline. For example, the American Society for Mechanical Engineers sets the criteria for mechanical engineers, the Institute of Electrical and Electronic Engineers (IEEE) for electrical engineering programs, and so on.

The purpose of the accreditation procedure is to help engineering schools and programs maintain high standards. The accreditation visitors will often go to bat for the departments and demand resources and attention from the university administration. Occasionally, they will withdraw accreditation from a program, and this can have devastating effects on the department or the university. Volunteers, engineers who give of their time and resources to assist in the betterment of engineering education and the continuation of the engineering profession, carry out the accreditation visits.

Later

You have never been able to decide about affirmative action, and you decide to ask Sarah's opinion. You're rather surprised to find that she's strongly in favor of it. "I had a real bad time at Purdue," she says, "and I wouldn't wish it on any woman today. All the faculty and nearly all the students were male, and they mostly treated me like I was Rosie the Riveter. Things would have been a lot different if there'd been more women—I felt really like I wasn't part of the main show. I had girlfriends in high school who could have had anything they wanted to, but they all ended up becoming secretaries, getting married and raising kids—no careers between the lot of them. If affirmative action's the way to get more women into engineering, I'm all for it."

You wonder if it is worth noting that affirmative action might also reduce the quality of engineering graduates, but you decide this is not a fight you want to pick, mainly because you are not sure of your facts or your values.

Box 10-8

Affirmative Action

Affirmative action began in the United States in the late 1960s as part of an effort to improve the economic position of members of groups that were perceived to be disadvantaged. The target group originally was African Americans, but over the years the range was extended to include women and various minorities. Affirmative action has always been controversial.

The early debate on affirmative action (or "reverse discrimination" as it was often known) focused on issues of compensation. African Americans, it was argued, were owed compensation because their ancestors had been captured and transported to America as slaves. The evils of slavery are apparent to all, and justice requires that the descendants of slaves be compensated for those evils. Critics replied that it is impossible to "compensate" people for harm done to their forebears—and that European Americans living today are not responsible for the harm done by their ancestors. Indeed, most of their ancestors were living in Europe at the time of slavery. Justice cannot be done either to or by dead people.

Partly in recognition of this argument, the emphasis shifted away from a "backward looking" approach to focus on the future. African Americans and other groups, it was argued, are disadvantaged *now* because of the continuing effects of past and, sometimes, existing discrimination. Gertrude Ezsorsky (1991), for instance, argued that even though women are now equal before the law, they continue to suffer from lack of self-respect and reduced standing and opportunities in the community, because of historic patterns of discrimination, unequal opportunities, and stereotyping. Thus, they are entitled to efforts from society to restore their lost equality, which is their right. In general, advocates of affirmative action argue that historical lack of educational opportunities, low expectations, underachievement, lack of inspiring roles models, and other factors continue to disadvantage various groups, thus entitling them to preferential treatment in areas such as admission to universities and training programs, financial assistance, job opportunities, and promotions. Many organizations respond with initiatives such as quotas for admission to medical schools and advanced training opportunities. Federal agencies give preference to firms owned by minorities and women in order to assist these groups to run successful businesses.

It must be emphasized that affirmative action was always intended to be only a temporary measure, to last only until disadvantaged groups ceased to be disadvantaged—that is, until members of these groups no longer had significantly worse rates of success in education, jobs, and business. In a 1997 issue of *Perspectives on the Professions*, several articles looked back on "25 years of affirmative action." Two themes dominate the material in the newsletter. First, the debate about the ethics of affirmative action continues. The second theme may be summed up in the phrase "Has it worked?" In the 1970s, people also asked questions about the effectiveness of affirmative action, but the debate lacked substance because no one had any idea whether affirmative action policies were actually going to deliver. Generally, predictions proceeded along ideological lines: People who believed that affirmative action was required by justice confidently predicted that it would work, and people who thought that it was not required by justice (many of them thought that it was quite *unjust*) equally confidently predicted that it wouldn't. For instance, some opponents claimed that students who were admitted to, say, engineering schools with inferior grades would struggle to keep up and would end up as second-rate engineers. Worse still, the public would assume that minority and female engineers were *all* affirmative action beneficiaries and wouldn't want to consult them; employers would doubt the validity of their credentials and wouldn't hire them.

Supporters of affirmative action denied all this, claiming that engineering schools could help bring affirmative action admission students up to standard by targeted assistance, as Ivan is providing (somewhat grudgingly, perhaps) at DTU. And as the public became used to dealing with successful minority and female engineers, they would judge professionals on their merits, regardless of their race or sex.

Now, 30 years on, we can look back and evaluate the success of affirmative action policies. The evidence seems to be that it has been very successful in

(continued)

> **Box 10-8** *(continued)*
>
>
>
> some areas, quite successful in others, and not at all successful in yet others. The military has been transformed from an organization totally dominated by white males to a diverse outfit with women and minorities represented at all levels of command. The stunning military success in the Gulf War shows that this transformation has occurred without any loss of combat ability.
>
> The military may be an atypical example—no other organizations have the ability to impose new structures on employees. Also, it may be argued that other factors are responsible for increased educational and professional success by women and minorities—civil rights legislation, for example.

Discussion Questions

10-1. Research one gift given by some person or corporation to your university. This may have resulted in the naming of a building or a program. Find out who gave the money, how much it was, and what the agreement was. If your development people stonewall you, the information can be found in past copies of your local newspapers, available in your library.

10-2. At the University of North Carolina at Chapel Hill, one of the large dormitory buildings is named after a fellow named Lenoir. The naming occurred in 1900 or so. Now it turns out that the guy was quite a scoundrel and a rabid segregationist. Many students want to change the name of the building, arguing that it is an insult to the university and its students to have a building named after such a scumbag. What do you think? Should they change the name of the building? Why or why not?

10-3. What exactly is so wrong about Coca-Cola sponsoring a lecture series at a School of the Environment? Maybe there is nothing immoral about this. What do you think?

10-4. Who are the big contributors to your college of engineering? Do they in any way affect the policies of engineering education at your institution?

References

deRubertis, K. 1994. "Discussion of 'Why Do Engineers Wear Black Hats?' by P. A. Vesilind." *Journal of Professional Issues in Engineering Education and Practice*, ASCE, 119 (1): 330–331.

Ezsorsky, G. 1991. *Racism and Justice: The Case for Affirmative Action*. Ithaca, NY: Cornell University Press.

Florman, S. 1972. *The Existentialist Pleasure of Engineering*. New York: St Martin's Press.

Schweitzer, A. 1933. *Out of My Life and Thought: An Autobiography*. New York: Henry Holt.

Vesilind, P. A. 1994. "Closure to 'Why Do Engineers Wear Black Hats?' by P. A. Vesilind." *Journal of Professional Issues in Engineering Education and Practice*, ASCE, 119 (1): 331–332.

11

Overseas work

Sunday, February 3

Girish has an enjoyable visit and is very impressed with your ability to deliver on what he obviously considers to be a promise to return a favor. He evidently thinks you have used your influence to get his kids—and probably his friends' kids—into De Tocqueville. This relationship, you decide, is definitely proving to be a useful one. And look at how much good you're doing—building bridges for international understanding, creating jobs here and in Genala, helping Pines, De Tocqueville, Girish, and of course, yourself and your family. Win-win!

The resort will take about 18 months to construct, and after discussions with Girish, Joe, Sarah, and the CEO of PKK, the Genalan construction company that is building the resort, it is decided that you and an assistant, Trish, will spend the first three months in Genala. After that, Trish will be based there for the duration of the project, and you'll make brief follow-up visits every two months or so. This is Trish's first overseas project. She's a competent engineer with a specialty in infrastructure, but she has also worked on building construction and it's a great career development opportunity for her.

The resort, which is named (unimaginatively, you think) the Hibiscus Retreat, is located on Luka, an undeveloped island just off the mainland. There is one very small town and several villages; the rest of the island is a nature preserve. The resort is designed to attract rich foreigners who want to stay in a luxurious resort but also have the chance to see wildlife and experience a real rain forest. It is located on a beautiful, palm-fringed beach that, you soon discover, has spectacular sunsets. Luka is connected to the mainland by a recently constructed bridge, which the local people welcomed because formerly they were dependent on an inefficient and rather unseaworthy ferry service to the mainland town where they made weekly trips to market their produce and handicrafts. You're glad that the project will help out the local people economically—they're poor even by Genalan standards. Locals have been hired for construction jobs, and once it opens, you learn, they will be hired as maids, gardeners, and cleaners. The resort will feature nightly cultural performances, in which local people will also participate.

Initially, you are surprised at the slow pace, by American standards, at which the infrastructure is proceeding. Much of the work is being done by hand, but you soon become accustomed to the sight of rows of men wielding shovels and lines of women carrying large baskets of rocks on their heads.

The project manager for PKK is Ah Chee. He's a very pleasant, friendly man, and you get along well with him. He often invites you and Trish to his house for a meal or drinks after work. Not long after you arrive, Trish meets Mario, a Filipino engineer on the project, and begins a whirlwind romance. You are of course often excluded from their activities, and this makes you especially glad to have Ah Chee's company. This evening, you are sitting on the beach with Ah Chee, Trish, and Mario, watching the sun go down and sharing a few beers and a bottle of Armagnac that Mario has somehow acquired. It's a small celebration: Trish has just sorted out a difficult problem in the drainage system that threatened to delay the project. Everyone is cheerful and relaxed. You don't have to work tomorrow: It's a public holiday devoted to a revered Hindu holy man who played a prominent role in the Genalan independence movement.

"This is such a fine place to work, surrounded by so much beauty," Mario says, gazing at Trish, who bears a striking resemblance to Geena Davis. Everyone makes noises of agreement, and Mario pours more Armagnac.

"I love it here too," you say, "and I like this project. It's doing a lot for the local people, too."

"They are hard workers," Ah Chee says, "and reliable too. This is a very good place to do business if you know what you are doing."

"As we do, I hope?" you say.

"Certainly. Some other companies, with less experience in the region perhaps, have however been quite unsuccessful."

"I can see that," Trish says. "For one thing, the water table's so variable, and those little volcanic tremors don't help."

"You are quite right, Trish. And also there is the problem that companies have because they are not attuned to the local culture."

"How's that?"

"Let me give you an example. Perhaps you know of Ulf, it's a Norwegian utility construction company."

We all nod. Ulf is a leader in hydroelectric construction in northern Europe and Africa.

"A few years ago, Ulf had a contract to build Genala's first hydro plant, in the mountains near Ramaya. They planned to hire about 25 local people to drive bulldozers and other equipment that they would rent from Ramaya companies, and they had budgeted to pay them $100 a week. Also, they needed some unskilled laborers, to whom they were planning to pay $75 a week. But the liaison man from the ministry of energy said they couldn't do that. Those are ridiculous wages, he said, at least three times too much for the drivers. Also, more of the work should be done by hand, so as to employ many more people as laborers, and they should be paid at the standard rate of $3 a day. The cost to Ulf would be the same, but there would be more jobs.

"The engineer in charge, a guy named Erik Johansen, was very angry. He said that in Norway the workers would be paid many times the rate that Ulf was originally planning to pay in Genala, never mind the new rates that the official was demanding they pay. His company recognized that wages and prices were much lower in Genala, and so they had decided what they thought were fair rates in the circumstances.

"The Genalan official would not hear of this. He said that they must follow local practice and pay the usual rate. Otherwise, they would create jealousy among other workers. Also, they could be encouraging labor unrest and there could be pressure on the government to increase the legal minimum wage. This could fuel inflation, and impede the country's development. There was quite an argument, and finally they reached some compromise, but no one really liked it."

"How much do we pay our laborers, then?" Trish asks.

"Three dollars fifty a day, which is somewhat above the market rate, plus a 10% cash bonus at the end of their employment on the project, for most of them that will be about an extra $75."

Trish is appalled. "Why, that's slave labor! In America they'd get at least five times that daily rate an hour!"

"But this is not America. Also, we provide free lunches, and medical care if an employee is injured on the job. We have set up a clinic for children's health, with some free prescription drugs, and a nurse to go around the villages. And you will have noticed the day-care center we provide, also free, for the workers who do not have family members available to take care of small children while their parents are at work. All this is much, much more than the law requires. Then, when the job is finished, in addition to the bonuses, the workers will be given all the hand and power tools they have used on the job. We are trying to train all of them in construction skills so that they do not have to spend the rest of their lives digging ditches with shovels and carrying around rocks."

"Yes," Mario says. "Then they can get skilled work on other projects, or maybe set themselves up as subcontractors, and everyone's better off."

"But in the meantime we are exploiting these people. It is their land and their country, after all, and the rich people are taking it away from them," Trish says with subdued anger. To head off a possible conflict, you ask Ah Chee to finish his story about the hydro dam.

Box 11-1

Environmental Racism

During the past few years, a new term has gained importance to engineers—"environmental racism." This idea originates from the empirical evidence that undesirable land uses such as incinerators, wastewater treatment plants, landfills, and the like are often sited in areas of a community with a high percentage of minorities. The undesirable land use therefore is unevenly and unjustly distributed, and this is interpreted as being racially motivated.

Although we do not want to support the uneven distribution of undesirable development, the uneven distribution of undesirable land use is much more likely due to economic factors—the land is simply less expensive in the poorer parts of town, and these areas are often the minority neighborhoods. Engineers and other decision makers are not necessarily racists, but they are beholden to the public to provide services at the least cost. But there are examples where racism comes close to the surface.

One such instance occurred in Chapel Hill, North Carolina. In the 1970s, a landfill was sited near a minority neighborhood because the land was inexpensive and the level of public opposition at the time of the siting process was low. The neighbors were assured that the landfill would last for only about ten years and then a new landfill would be constructed in another part of town. When the ten years passed, and a new landfill became necessary, the town decided to enlarge the existing landfill because the problem of siting new landfills had become politically too difficult. Then in 1998, the town council decided to buy more land contiguous to the existing landfill to extend the life of the landfill more than 30 years. The town officials told the people in the minority neighborhood that the promises to move the landfill to another location could not be binding because these promises were made by a previous administration.

There is clear injustice here, and breaking promises made by previous local administrations is blatantly immoral. Is this, however, a case of racism, or is it a case of gutless politicians simply breaking promises?

Whatever the reason for injustices in the imposition of societal costs such as undesirable land use, the concept of "environmental racism" has added another variable into the decision making of engineers and governmental agencies. There is no ethical justification for the unequal treatment of citizens, regardless of race or economic standing. The recognition of such inequities has made it quite clear that economics will no longer be the sole criterion governing the location of undesirable land use and the distribution of societal costs: and this is how it should be.

"There were so many problems," he says. "From the beginning, all kinds of things were disappearing—tools, small pieces of equipment, wheelbarrows, anything that could be moved by hand. Johansen was in despair. He was certain that the workers must be stealing the tools, and he hired some people as security, but they never caught anyone. He raised the problem at a meeting of the project engineers, and one of the engineers, a New Zealander, Ben McDougal is his name, in fact he is the same Ben McDougal who started with us yesterday, asked him how much tea money he was paying. Erik had no idea what he was talking about, so Ben had to explain it to him.[1] I think you can see him surfing out there. Maybe he will come and have a drink with us when he has finished surfing—it will be dark quite soon."

Sure enough, Ben catches one more wave and then wanders up the beach carrying his board. "Ben, come and join us for a beer!" Ah Chee calls out. Mario hands Ben a beer and Ah Chee introduces everyone. Ben studies Trish carefully. "I recognize you: Haven't we met before?"

Trish lived in New Zealand for a while and recognizes this as a standard Kiwi pickup line. "I don't think so."

"Funny, you look really familiar."

Ah Chee says, "Ben, I was just telling our friends here about Johansen's problem with the tea money."

"Yeah, Erik the Viking, we used to call him, he was such a crusader."

"I didn't know there were any Viking crusaders," you say.

"Course there weren't, but he would have been if he'd been in the right place at the right time. It was just a joke. Like, Erik's always up on the moral high horse, wanting to convert the infidels to his way of thinking: universal human rights, et cetera. So, when I explain about the tea money and stuff he goes all pompous and he's like, 'I shall not pay my workers not to steal! Do these people have no respect for property?'"

"Sounds like a pretty corrupt system to me," Trish says.

"Unbelievably! And you're wearing an Expozay bikini that cost you at least two hundred bucks."

"Very perceptive of you, but so what?"

"Yeah, sure. Not that perceptive, my girlfriend's a manager for Expozay back home, they export heaps of stuff to the U.S. But you see that big coil of copper wire over there? How many days do you think our laborers have to work to buy that? So eventually, Erik agrees to another compromise, pay the local headman $1000 a week as a security consultant, and the thieving stops, just like that. Course, the headman has to spread it around, so everyone gets something. And like it's only about 75 grand on a five-million-dollar project anyway."

"I don't like all this 'When in Rome' stuff!" Trish exclaims. You've heard this before. You've *said* it before.

"No, and I suppose not when in Auckland, when in Oshkosh."

"What do you mean?"

"Suppose a Genalan company does work in Oshkosh, if there really is such a place, what a name. They do it your way. Here, you do it their way. Simple, eh?"

"But that's different. We pay proper wages and we don't pay workers not to steal."

"Oh. How do you deal with theft? You do have theft in America?"

---

[1] *Tea money* is a term for payments that contractors in Asia often make to local people to ensure that the site is not a target for theft and to ensure the quality of locally sourced labor. In rural areas, the payment is made to the village headman. It is not illegal under the Foreign Corrupt Practices Act. The narrative goes on to explore the ethics of tea money.

"Of course we do. So we hire security staff and anyone who gets caught stealing is fired. If it's serious, we bring in the police."

"Oh."

Apparently oblivious of Trish's mounting anger, Ben goes on: "My brother was working on a project in Bangkok once, and they were pouring the foundation. The first load of concrete arrives four hours late and it's already halfway set. You've heard about Bangkok traffic? So this cop walks over and he's like, I can fix this up for you, I'll sit in the cab with the next load and make sure it gets to the site on time. So my brother gives him some baht,[2] but the load is still two hours late and again it's no good—evidently the cop didn't know how quickly it sets. Then the cop offers to guarantee delivery but at some cost. With a couple of traffic cops on motorcycles, lights flashing, sirens blaring they get the concrete to the site. Cool, eh?"

There's a pause. You're getting tired of the conversation and you announce that you're going for a swim. As you head off you hear Ben saying to Trish, "I know who you remind me of, that chick in that movie about girls' baseball teams in World War Two, not Madonna, the other one."

"That . . . *chick* . . . ?"

"Bye-bye, Miss American Pie," you think, suddenly struck by the fact that today is the anniversary of "the day the music died."[3]

Monday, February 4

Next day you run into Ben at lunch, and ask him to tell you more about the hydro project.

"What finally did it," he says, "one of the workers started distributing antigovernment pamphlets demanding justice, respect for rights, political freedom. The government guy found out and told the Viking he had to fire the worker immediately. Of course Erik refused, said he was only exercising his democratic human rights, blah blah.

Box 11-2

Human Rights II

The modern idea of universal human rights dates back only around 300 years, when it was referred to as "natural rights." Key texts include the writings of English philosophers John Locke (1632–1704) and Thomas Paine (1737–1809) and French philosopher Charles, Baron de Montesquieu (1689–1755). The works of these writers were very influential in revolutionary movements, especially in France and America—the United States Constitution was strongly influenced by their ideas.

The original doctrine of natural rights was based on Christian principles—Locke, for instance, believed that the rights to life, liberty, and property were part of natural law, which is ordained by God and which all humans are able to understand by virtue of their God-given reason. The U.S. Constitution states: "All men are endowed by their Creator with the rights to life, liberty, and the pursuit of happiness." Jefferson substituted "pursuit of happiness" for "property" to get around the
(*continued*)

---

[2] We would call it a "tip."

[3] In Don McLean's iconic "American Pie," the day the music died was February 3, 1959, when Buddy Holly, Ritchie Valens, and The Big Bopper died in a plane crash in a snowstorm, symbolizing for McLean the end of rock and roll.

Box 11-2 (continued)

sticky question of slavery. How could all men be created equal if some of them were the property of others?

Modern statements of universal rights, including constitutions, human rights legislation, and the writings of political theorists and philosophers, are usually presented in secular terms, but the content of these documents—the fundamental rights that, it is claimed, we all have—does not vary greatly between religious and secular documents.

The roots of the ideal of universal equality that underlies natural rights theory in European civilization go back 2000 years to the ideas of the Roman Stoic philosophers such as Seneca (4 B.C.–A.D. 65) and to the Judaeo-Christian tradition. For instance, the story of the Good Samaritan is profoundly egalitarian. Much earlier, the Lord Buddha taught "universal compassion," and he saw his mission as being "to work for the well-being of the many, out of compassion and for the benefit of all" (deSilva 1998). But the idea of rights as an entitlement that each person has, regardless of the circumstances, is found only in post-seventeenth-century political writings.

Philosophers and legal political theorists who advocate human rights usually appeal to the following:

- Universality: Everyone has human rights.
- Innateness: Human rights do not have to be earned.
- Inalienability: Human rights cannot legitimately be taken away.

Recognizing that few if any states fully protect what are claimed to be human rights, theorists distinguish between *moral* rights—the rights that they believe everyone ought to have—and *legal* or *civil* rights—the rights that are in fact protected by law. Because even legal rights are not always in practice protected, human rights are often seen as ideal statements by which to measure our institutions and as goals to which to aspire. This is made explicit in the 1996 South African Constitution, which states that everyone has a right to have adequate housing and a healthy environment, but acknowledges that this is a long-term goal, not an immediate entitlement. So far as we know, this is the only national constitution that is honest enough to state that people have rights that cannot be achieved immediately.

The Western concept of human rights is highly individualistic. The core notion is that rights are a special moral entitlement that must be upheld, regardless of any harm to society that might flow from their exercise. By contrast, in many non-Western countries, including those that include rights in their constitutions, equal if not greater emphasis is placed on societal needs such as national development.

Thus, in many parts of the world, especially former European colonies in Asia and Africa, Western countries that condemn countries such as Nigeria, Burma, Indonesia, and China for human rights violations are seen as attempting to impose an alien culture, to globalize Western values, and—especially in the case of former colonial powers—as trying to reassert the control they once had over their now-independent former colonies.

Even in the United States, most people probably do not believe that rights such as property rights or the right to bear arms are *absolute* in the sense that they must be protected regardless of all societal goals. Thus, most Americans appear to accept that the government is entitled to take a proportion of everyone's income in the form of taxation, to restrict access to firearms, and to compulsorily purchase land for highways—all of which are limitations on property rights. But an increasing number of Americans are suspicious of government and believe that government powers should be sharply reduced because they violate natural rights—that taxation, for instance, violates property rights by taking people's property without their consent for purposes that they do not support. Key writers in support of this view include Robert Nozick (1974), Ayn Rand (1905–1982), and Frederick Hayek (1899–1992).

"Next thing you know there's another government official with a whole squad of guys in uniform and carrying Kashas, who tell him the government has canceled the project and he and all his crew have to be off the site in four hours and on the first plane out of Ramaya. When the Viking protests that Ulf has a contract and everything, and how dare they, et cetera, the official whips out a copy of the contract and points to a clause that Ulf agrees to respect the local culture and not do anything that will undermine the traditional fabric of Genalan society. I guess Erik thought that only meant things like wedding ceremonies and religious festivals and art and stuff."

"What happened to the project?"

"Oh, soon as I see what's going on, I get on the phone to a mate of mine back home, CEO of an engineering firm that specializes in hydro. He calls the Genalan embassy in Wellington and a week later Aotearoa Construction is hard at work finishing off the project."

"Ayuh what?"

"Aotearoa. It's the Maori name for New Zealand, and it's a tribally owned company. They know all about respecting local cultures because they had such a hard time maintaining their own. So then there's no problems, and right now the city of Ramaya is humming along on power from that dam."

"Sounds like you go along with political repression, then," you say, somewhat stiffly. "Don't you believe in democracy?"

"Ooh yes, let the Genalans have a democracy like yours, then they can have a president who only knows how to breathe out and didn't have sex with that woman."

Box 11-3

Politicians and Their Reputations

You may very well feel that this is a cheap shot and unfair. We should also tell you that a reviewer of our manuscript thought that in mentioning this, we were disrespectful of the presidency.

It is indeed a cheap shot, but it does not misrepresent statements made by former President Clinton. Ben is expressing a view that is widespread in America, as well as the rest of the world, that democratic institutions have become corrupted because the people who occupy positions of power within them are less ethical than they should be. Perhaps it is unfair to judge political systems by reference to the character of their leaders. But we do this all the time. Iraq, Israel, Iran, Malaysia, and Indonesia are or have been evaluated by the performances of leaders such as Saddam Hussein, Benjamin Netanyahu, the late Ayatollah Khomeini, Muhammad Mahathir, and Suharto. The message of this book is that every engineer ought to be an ethical person. The president of the United States of America, today, is the most powerful person in the history of the world. All the more reason to insist that the president, also, be an ethical person.

Also, recall Bok's argument that when politicians are caught lying, the public's faith and trust in government and democratic procedures is weakened. Bok was writing in 1979, well before the tenure of president Clinton.

Discussion Questions

11-1. The standard practice in American restaurants is to tip the waitperson about 15 to 20% of the cost of the meal. In Europe, you don't tip because a 15% "service charge" is added to your bill, to be distributed to staff. In New Zealand, waitpersons are paid a living wage and do not expect tips, but many people (including ethicist Alastair) leave a few dollars on the table as a reward for excellent service. Does that mean that in the United States we are operating under a bribery system not unlike the need for "tea money" in many other parts of the world? Is there a difference between tipping a waitperson and giving money to a foreman for distributing to the workers? Are both or either immoral acts? Present your case.

11-2. Suppose the Congress of the United States passed, and the president signed, a law requiring all public works contracts to pay 5% off the top to the governors of each state, to be used for other public works projects. Would this be acceptable morally? Why or why not? How would this differ from paying (under the table) 5% of the contract price to the president of a small and poor county in which you are constructing a public works project, funded by the U.S. government?

11-3. Should companies that are operating in low-wage economies overseas pay local workers at local rates (and provide local conditions of employment), or should they pay the same rates (discounted for local cost-of-living differences) as in their home country?

11-4. Are there any universal human rights? On what could such rights be based? Or is the human rights movement no more than an attempt to impose "Western" values on other cultures?

References

deSilva, P. 1998. *Environmental Philosophy and Ethics in Buddhism*. New York: St. Martin's Press, pp. 85 and 74.

Hayek, F. 1944. *The Road to Serfdom*. Chicago: University of Chicago Press.

Nozick, R. 1974. *Anarchy, State, and Utopia*. New York: Basic Books.

Rand, A. 1943. *The Fountainhead*. New York: Bobbs-Merrill.

12

Uphold the honor and dignity

Tuesday, February 5

You're working at your desk when you hear a loud yell from Trish's office next door, followed by "Get your stinky paws off me, you damned dirty ape!" Either Charlton Heston's had a nasty accident, or something has happened to Trish. You rush next door to find a furious Trish yelling at a large and, you have to admit, somewhat simian man in coveralls. You've seen him around the site: He's a construction foreman, an Australian named Dave. He's grinning and shrugging as Trish continues to berate him angrily.

"Hey, what's up?"

"This neanderthal just propositioned me, and when I told him to get lost he grabbed my butt!"

You shoot a questioning look at Dave.

"Aw, yeah, I was just horsing around, y'know? Didn't mean anything, really."

Trish says, "No you weren't! You sexually assaulted me and you're gonna be *real* sorry!"

Ah Chee, whose office is just down the hall, arrives. "I heard shouting. Is there some problem?"

Trish explains, in more temperate language, what has happened. Ah Chee asks, "Dave, is this true?"

"Sort of. Well yeah, I suppose. But I didn't mean any harm, it was just fun, like."

Ah Chee looks sternly at him. "I will not have employees creating a disturbance on the job. You will treat each other with *respect*. Dave, you will apologize to Trish, and you will promise never again to engage in such unruly behavior while you are working for this company."

Dave glares at Ah Chee, turns to Trish and, with some effort, mumbles, "Jeez, Trish, I'm really sorry, and I won't do it again."

Box 12-1

Manners

The authors of this book disagree about the universal status of ethics, but they agree that manners are culturally relative. Only a cultural imperialist wants to impose his or her society's manners on other societies. Manners are just a matter of polite behavior and "When in Rome" (see Box 9-3) rules, legitimately.

When I (ethicist Alastair) first visited America, I was amazed that students, even Ph.D. students, and secretaries called me "Dr. Gunn or "Professor Gunn." In New Zealand, everyone is on first-name terms, and even my freshman students (and the students in a high

(*continued*)

Box 12-1 (continued)

school course that I teach) call me "Alastair" or "Al." Of course, I prefer to be called by my first name. That is how my society operates, and I am comfortable with it because I am used to it. But I don't expect American students to do it. On the other hand, in Norway, where engineer Aarne did his post doc, people call each other only by their last name—no Mr. or Dr. or any other title. From secretaries to the head of the institute, all salutations are by the last name. This would be considered impolite in America, and probably just weird in New Zealand.

We guess that there is no such thing as inherently good manners. Consider the socially acceptable rules of driving; gossip, kissing. Just so it works.

Manners are not unimportant, though. They involve courtesy, respect, and caring for each other. They are part of the "social glue." Manners are what the African American community in southern United States calls "home training," a delightful expression that succinctly captures the heart of manners. Manners are even part of national identity. Java, in Indonesia, is a very polite society. Unruly children are said to be "not yet Javanese."

Although manners are a matter of convention, they are also often connected to important social requirements. In many societies, we greet strangers with a handshake. This custom derives from a symbolic indication that we do not intend to draw our swords to attack the stranger. Also, in Western societies men and women shake hands on being introduced, but in conservative Islamic societies they don't, because they are more conservative in matters of sexual morality than are Western societies, and physical contact—minimal though a handshake might be—between men and women therefore has a different meaning.

In highly stratified societies, manners have been used to maintain the status of elite groups, making it difficult for members of the lower class to behave "properly"—which knife or fork do I use? But even in our mostly egalitarian cultures, we need other members of our society with whom we interact to be predictable and easy to get on with. Hence, manners.

As we note in Box 4–5, engineers work together. In engineering, we resolve to have good manners by following many of the requirements in a typical engineering code of ethics. For example, Guideline 3e of the Code of Ethics of the American Society of Civil Engineers states:

> 3*e*. Engineers shall be dignified and modest in explaining their work and merit, and will avoid any act tending to promote their own interests at the expense of the integrity, honor and dignity of the profession.

Statements such as this are simply codified good manners. There are no further explanations as to what exactly is meant by "dignified and modest." We all know, however, when someone is *not* dignified and modest, and polite engineering society frowns on such behavior.

"Well, that is settled, then. And now, let us all get back to work and forget all about this unpleasant incident. Dave exhibited bad manners and has apologized."

"No way," Trish says. "What he did was sexual harassment and assault, and I want something done."

"Dave, go back to work. I will talk to you later, and please close the door behind you," Ah Chee says. After he's left, Ah Chee continues, "Now, Trish, sexual harassment is not an offence in Genalan law."

"It damned well should be! And anyway it's against company policy. Look what it says on the letterhead—'PKK is proud to be an equal opportunity employer.'"

"We are indeed an equal opportunity employer! We do not discriminate against potential employees on the grounds of race, gender, religion, and so on, as you will agree. Also, we have what in America you call a 'family friendly' policy, our day care centers and so on."

"I acknowledge all that, but if a woman can't do her job without being sexually propositioned and assaulted by apes like this one, then you're not really practicing EEO. I want that bastard fired!"

"I will not do that. It is one quite small event. Also, he is an excellent and experienced worker and he gets on very well with his team. This is the first time, to my knowledge, that he has misbehaved, and I am sure you will not have any more problems with him."

"Well, at least I want him disciplined, and warned that if anything like this happens again, he'll be fired."

Ah Chee says emphatically, "What I decide to do about Dave is a matter between Dave and myself. Let's get back to work!" and he walks out.

"Wow, Chris, thanks for all your support!" Trish snaps, and she too walks out of the office.

Box 12-2

Workplace Harassment

Chris chose to not enter the discussion, perhaps because there appeared to be no responsibility for doing so. Still, Trish would have expected some support from her boss. Should Chris have taken her side, or at least supported her in some way?

Trish seems to have a legitimate gripe. Dave's action is a criminal assault in many countries, although it is not criminal in Genala. Some people would say that it isn't a serious assault, and certainly she hasn't suffered any bodily harm. But her bodily integrity has been invaded, and she is justifiably angry and distressed. She may also be afraid that Dave—a physically intimidating man—may proposition her again, or worse.

Let's agree that everyone has a right to a safe and nonthreatening workplace environment, and that they're entitled to be free of all forms of harassment on the job. In the United States, and many other jurisdictions, it is a legal right to be protected from such behavior. Furthermore, many organizations explicitly prohibit it in their employee code of conduct. Under U.S. law, as we discuss earlier, the "hire and fire at will" rule allows private-sector employers to dismiss any employee at any time, unless they are protected by state law or the provisions of their contract. But even in countries with tough employee protection legislation such as Scandinavian countries, Dave could be in serious trouble, and if he repeated his behavior he could be liable for dismissal.

Many firms around the world have adopted a conciliatory approach to situations such as the one in the narrative—for instance, by hiring a mediator to deal with such cases. The mediator's job is to receive and investigate complaints such as harassment, and to arrive at a resolution that is satisfactory to all parties. This is what Ah Chee tried to do. He heard both parties' versions of the event, and they agreed about what happened. He made Dave apologize and promise not to do it again, and he hoped that Trish would be prepared to let the matter rest. But she isn't—and you may think that she is quite right.

Later that Evening

You and Ben are having a beer and a plate of spicy noodles at a makeshift restaurant set up near the site by an enterprising local family, discussing the prospects of yet another Green Bay Super Bowl triumph, when Trish arrives. She is looking hot and angry and ignores you. As she's getting her drink, you leave your table and go up to her. "What's up, Trish?"

She glares at you.

"Hey, I know you're mad at me, but I'm not Dave's boss. It was up to Ah Chee to sort it out, he's the boss."

" I suppose," she says reluctantly. "But you could have said *something*."

"I did, to Ah Chee this evening. I said that how he treated his team was his business, but how they treated my team was mine. He said he understood and there wouldn't be any more trouble like that, he personally guaranteed it. So, why don't you come and join me and Ben?"

She looks around, but the other tables are occupied so she nods.

"Hi, Trish, how's it going?"

"OK I guess. Ben, that bastard Dave . . ."

"Yeah, bad news, eh? Heard he'd tried it on with a couple of other women before. Mario should pop him one."

"Mario?"

"I thought you two were an item?"

"Highly unlikely. He said it was nothing important and I shouldn't take it so seriously."

Discussion Questions

12-1. In the scenario above, what should Trish do next? Remember that she is in a country where sexual harassment is not illegal. But she is working for an American corporation, and sexual harassment is certainly illegal in the United States.

12-2. Alastair's draft of this scene had Ben give Dave a black eye as retribution for his boorish behavior. Aarne disagreed. We recognize that having Kiwi Ben give Aussie Dave a black eye is an attractive story line, but on the other hand the act might be construed as supporting violent vigilante tactics in resolving social problems. Do you think that in our story Ben should have laid Dave out? Why?

12-3. In some societies, Dave's actions are not only not criminal, but they are considered to be a compliment. If you are a woman, which society would you prefer to live in? Why? If you are man, which society is more appealing to you? Why?

13

Faithful agents

Wednesday, July 12

The rest of your time in Genala is uneventful, thankfully. In fact, the next six months of your life is fairly routine. Pines continues to do lots of business. Joe hires a number of bright young and not-so-young engineers to replace the people who died in the crash, including a Genalan, a recent graduate from MIT with an excellent academic record and glowing recommendations from the school.

After your brief period working at project level, you find yourself back in your old, managerial position, though with an increasing focus on overseas operations. Several times you wonder about applying for other jobs. You've rather lost faith in Joe and Sarah, and relations between you are not as friendly as they used to be—polite but a little cool. You pay several more visits to Genala, including the grand opening ceremony for the resort, performed by the minister for the environment herself. The night before, Ah Chee holds a big party for the entire remaining workforce, though most of the expats have already left for other projects. Mario is nowhere to be seen, but you notice that Ben and Trish spend most of the evening together.

You are asked to present a talk at the national meeting of ASCE in Cincinnati. Your talk on finite element analysis is well received, and a number of people compliment you on some original ideas you presented. At one of the many closing cocktail parties that are ubiquitous at a professional conference, an engineer from Walker Engineering, Paul Marshall, introduces himself to you. Walker is a a large corporation, Bechtel-sized but not as well known to the public, because it does business mostly through subsidiaries.

"I liked your presentation very much. You must have done a great deal of work in that area."

"Some, yes." Not all of it successful.

You talk some more and exchange business cards.

Wednesday, July 19

Paul Marshall contacts you—he is to be in town soon and wonders if you would like to meet for dinner and talk about some matters of mutual interest. "Oh, and please ask your partner along too."

"I'm not sure that Alex is free, but I'll get back to you. I'll certainly be there."

You hang up the phone.

"Looks like you're being headhunted," Alex says.

"It certainly does. Wanna come along for the ride?"

"Sure, but don't expect me to move to Delaware." Like many corporations, Walker is registered as a company in Delaware, though unlike most of them it actually has its head office there.

"Me either, but it's not very likely." You explain about the subsidiaries.

"I remember now, they have names like Pacific Structures, Tucson Engineering, Maine Environmental Services."

"How on earth do you know that?"

"Oh, it's their marketing strategy, textbook case. Clients see Acme Worldwide Engineering, it's just another anonymous international company, but call yourself Green Bay Engineering Services or Rocky Mountain High-Rise and you sound local, clients can identify with that."

"I don't want to move to Green Bay either."

"Why not? Then you could support the Packers and freeze your tush off at Lambeau Field. Anyway I made that one up."

"So, let's just see what they're offering. You're smarter than me on this."

Later that same day

You have a pleasant evening with Paul and his wife, Caroline, who is a marketing consultant. They're from England but have lived in the United States for many years—they're quite a bit older than you. They're both tall, blond, and conventionally dressed—fitting your stereotype of the British professional couple. She's even wearing pearls.

As the evening goes on, Caroline begins to engage Alex in a detailed discussion about marketing theory, evidently a ploy to enable Paul to gain your full attention, since, like engineers generally, you are not interested in marketing, let alone its theories.

British or not, Paul's approach seems quite straightforward. "You must be finding life a bit quiet at Pines."

"Why do you say that?"

"Oh, well, you were saying that you sometimes wished that you could get out of the office more, attend fewer meetings, and put on a hard hat, that sort of thing."

"Right. Though," you add loyally, "Pines is a good firm to work for."

"Certainly. Of course, larger firms do have rather more to offer, firms such as Walker, know what I mean? By the way, did I give you my card?"

You both know that he did, but you take it anyway. You study it carefully. On the earlier card, Paul was described as "Senior Design Consultant," a somewhat vague position, you'd thought at the time, but still, they keep changing the names. According to this card, he is "Executive Vice President Recruitment."

"You had a promotion in the last few weeks?"

"Ah, no. I must have given you an old card, silly of me. Walker is of course an expanding organization and always looking for new talent. At the moment we're keen to recruit middle-level people with the potential to fill relatively senior positions in our umbrella group . . . I presume you're familiar with our rather unusual structure?"

You don't want to look like an easy catch, so you say, "Yes, and it's very interesting," and you move on to another topic. But as you leave, Paul takes you aside and says, quietly but with conviction. "You must have gathered that we're interested in you. We've done some checking, and we're prepared to offer you an attractive package with excellent prospects. And no, you won't have to relocate to Delaware, all we have there is our finance division. You could live here and commute to our regional office in Atlanta, and telecommute some days if you wish. We can work something out, I'm sure. My own office is in Worcester, Massachusetts"—light grimace—"but I certainly don't *live* there.

"You are making this very difficult for me."

"Exactly my plan. Look, if you are undecided, give us a try. Why don't you start slowly, work for Walker a couple of days a week, just to see how it is. Get to know the people. You don't have to say anything to Joe or the other chaps at Pines. They will never know. We'll put you on the payroll at part-time, say two days a week, and you do as much as you can. If you don't like it, you can go back to Pines and nobody will have to know. What do you say?"

"I'll think about it." But what you're thinking is that you do not like this one bit. For a while there, you think, he had me on the ropes, and then he made the decision easy for me. No respectable firm would make such an offer. Engineers cannot have loyalty to two different firms at the same time. You say goodnight and vow not to ever get involved with Walker Engineering.

That evening

"Take it," Alex says, "You're not having a great time at Pines these days."

"The job's 200 miles away."

"Sure, but you can probably work from home most Mondays and Fridays. Rent a place in Atlanta and you can go to midweek games."

"I don't want to be involved with a company like that. Joe has been very good to me and I owe him."

Box 13-1

Loyalty to an Employer

Loyalty is a basic moral value. If we are loyal to each other, we all benefit, even if loyalty seems to be contrary to one's selfish interests. One way of demonstrating this assertion is to use what is known as the "prisoner dilemma." The story is that you and your (loyal) friend have been unjustly imprisoned, and each of you is asked to rat on the other, thus saving your own skin. You are being held in separate cells so you cannot communicate with each other. The problem can be illustrated like this:

		You accuse your friend	
		Yes	No
Your friend accuses you	Yes	You both get 5 years	Your friend goes free, you get 20 years
	No	You go free, your friend gets 20 years	You both get 1 year

(continued)

Box 13-1 (continued)

You have two choices. You can accuse or not accuse your friend (falsely). If your friend accuses you, then it is to your benefit to accuse him. (5 years vs. 20 years). If your friend does not accuse you, it is still to your benefit to accuse him (you go free vs. 1 year). In other words, whatever your friend does, it seems that the action that will benefit you most is a selfish action, to accuse him.

But the problem is that your friend thinks the same way! He also knows that regardless of what you do, it is to his benefit to accuse you, in which case you both lose. But if you and your friend feel a strong bond of loyalty toward each other and would not falsely accuse each other, then both of you win. That is, an action that seems at first to be contrary to your selfish interests is actually to your benefit.

Loyalty to a firm works the same way. If the firm is loyal to you, and you are loyal to the firm, you both benefit. But the assumption here is that the firm *deserves* your loyalty (and that you deserve the loyalty of the firm).

There are two types of loyalty: 1) loyalty to a just cause or person, and 2) loyalty to a person or cause that may not be moral or just. The latter is often called *blind* loyalty, or loyalty that exists whether or not it is deserved. Recent examples of blind loyalty are the employees (particularly accountants) who encouraged, by their silence, the unethical and illegal conduct of company executives that led eventually to the downfall of major firms such as Enron and WorldCom. They thought they were being loyal to their own accounting firms by not saying anything, but these firms did not deserve their loyalty.

The most important development in defining the moral and legal limits of loyalty was the trial of Nazi war criminals following World War II. The generals and other leaders were accused of horrific crimes against humanity. Their defense was that they were simply following orders. They believed that they were not guilty of these crimes because the orders had come from higher up. The judges at these trials, held in the German city of Nuremberg, firmly squashed this defense and ruled that loyalty is moral only if the cause is just and that "following orders" is not a legitimate defense. The defendants ought to have understood that what they were doing was criminal at the extreme, and ought to have refused to participate.

American universities certainly encourage loyalty from their students, faculty, and alumni, and we all agree that this cause is just. Of greater concern is the loyalty to one's friends when the group is engaged in immoral (or just stupid) behavior. Hazing at a fraternity, for example, is both illegal and immoral, and yet it is very difficult for one person to stand up and say that they will not participate in hazing. The loyalty to the group, and the fear of exclusion, is very strong, and it takes a special person to resist such pressures, and an even more special person to convince others to not participate in such activities. Loyalty to friends engaged in immoral behavior is blind loyalty, and is not condoned or excused.

Thursday, July 20

In the morning, Joe calls you into his office.

Without preamble, he says, "Hear you had an offer from Walker."

"Say what?"

"You heard. You want to work for them?"

You've no idea how he knows about the offer, but he certainly won't tell you and there's no point in denying it.

"I like it here and Pines has been an excellent employer."

"Not going to rush into anything I hope." A statement, not a question.

"Hell, no."

"What with the overseas side of the business expanding so rapidly, all thanks to you, and don't think I'm not grateful, I think you do have a great future with Pines, and I do like to reward loyalty."

"Well, thank you."

"We'll see what we can do."

You look each other in the eye and shake hands. There is a sense that your relationship has changed forever.

That night

That night you dream—not for the first time—that Alex, James, and Laura are standing under the overhang at the entrance to the Asmara, waiting for the hotel courtesy bus to the airport. You're across the street in a technical bookstore. The weather is horrible, high winds and rain, but you really want to pick up the first issue of a new computer magazine to read on the plane—if it ever takes off in this storm. As you start to cross the road to the hotel, you hear a terrible grinding noise from the direction of the overhang and you run wildly across the street, ignoring the traffic and yelling, "Move it, run, now!" at the top of your voice. But no one notices that anything is going on. Just as a large crack appears in the overhang, and masonry begins to topple, you wake up, usually to the reassuring presence of Alex beside you. But this time, you're on your own in the Atlanta Marriott. Try as you might, you can't get back to sleep.

Discussion Questions

13-1. We argue above that an employee has the moral obligation to follow morally acceptable orders from superiors in a corporation. But morally acceptable to whom? Suppose an employee believes that God has decreed that men are superior to women and that it is morally unacceptable to take orders from a woman. The employee believes he is acting morally. Should such action then be allowed? Why or why not?

13-2. Professional baseball players, some of whom are part of this story, expect the fans of their club to adopt them and show their support by paying high ticket prices to come to the ball games. And yet there seems to be no reciprocal loyalty. With some exceptions (Hank Aaron, Paul Molitor, and Sammy Sosa come to mind), most highly paid ballplayers will leave their current team as soon as someone (often the New York Yankees) offers them more money. If they expect loyalty from their fans, shouldn't they show some loyalty in return?

13-3. What do you think it means to be loyal to your college or university? Can such loyalty ever have ethical dimensions?

14

Avoid conflicts of interest

Friday, July 21

The next morning, Joe calls you into his office. Sarah is there. Without preamble, Joe says, "Want you to go to Atlanta next Thursday and negotiate a contract with P2O."

"As you know," Sarah says, "P2O is a large consulting engineering and construction company. The engineer you'll be talking to says she knows you, you worked together on a project once in Dallas. Gina Robbins. Her husband's a nice guy too, he went to school at Georgia Tech with my eldest, and . . ."

"Husband's name's Sherman," Joe interrupts. "Bet he's popular in Atlanta, haw haw. Heard he tried calling himself Lee Jackson at first but they found out his real name."

Joe, not a natural kidder, is obviously trying to get back on friendly terms; you go along with it.

"I remember Gina from our student days at Lehigh. Good engineer."

"Should be able to fix it up by Friday afternoon," he continues. So, stay for the weekend—on the company, room and board. Watch a ball game or two. P2O has a corporate hospitality box at Turner Field, so you'll have a good view of the game."

Hank Aaron Field, it should have been, you're thinking, but you know that Joe is a great admirer of the Braves owner and media mogul.

"Anyway, I looked up the schedule, there's the last of three against the Mets on Friday and Cincinnati Saturday. Maybe you'll get to meet Marge, hey hey, hope you're not Jewish."

You're dumbstruck by this remark. Every baseball fan knows all about Cincinnati owner Marge Schott's many gaffes such as her public praise of Hitler's organizational abilities. But how in hell did Joe dope this out? He hates baseball. In fact, for some reason, he's a *cricket* fan—he even subscribes to a cable service that provides coverage of un-American sports around the world so he can watch it. He really must be making an effort.

Sarah proceeds to brief you extensively on the job, which is a huge port extension at Galveston. P2O is designing and building the project, and Pines is subcontracting to design part of the onshore work. You're surprised that P2O isn't doing the design work in-house, but Sarah explains that the company has landed a number of major projects at the same time, and doesn't have the capacity to do all the work.

Thursday, July 27

Gina picks you up at the Atlanta airport in a new orange Porsche, a car and color that, you recall with a smile, Alex holds in complete contempt. Gina is tall and blond, and unlike her student days, is wearing a severe tailored black suit, no doubt befitting her corporate role.

You chat away about old times and more recent ones as Gina fights her way through the traffic along I-85 to the P2O office on Peachtree, near the state capitol. It's a fine day, and the golden dome roof is dazzling. As Joe predicted (and probably arranged, you guess), she invites you to be the company's guest at either or both ball games. You accept enthusiastically, what else?

Friday, July 28

Business is concluded amicably and to mutual advantage by early Friday afternoon. You spend the afternoon wandering around Atlanta's revitalized downtown.

You remember the first time you were in Atlanta, when you were still an engineering student, and visiting your Great Aunt Meg who lived out near Emory University. Despite her age, at least 100 years in your estimation, she'd encouraged you to go out and have fun.

"Is it safe there for me on my own, I mean . . ."

"Nowhere's safe just like that, honey. Y'all just look safe, you *are* safe."

Box 14-1

Safety

How safe is "safe"?

Nothing of course is totally safe. Whatever you do, some harm might come to you. You are "safe" in bed and your house gets hit by a light plane. You are "safe" in your kitchen and the hot water heater explodes. You are "safe" in a barn and it gets hit by lightning.

Engineers are often put into the position of having to design to some level of safety. For example, in highway engineering the amount of money we can use to design a totally safe highway is infinite. We can also spend an infinite amount of money to take out all contaminants from a public water supply, but society cannot afford such costs, and the engineer has to decide what is safe enough.

One definition of safety might be that a thing is safe if the risks are judged to be acceptable. Thus, skydiving is safe for the people who choose to accept the risks associated with this sport, but the majority of the public would consider skydiving unsafe because they would not accept this risk. Unfortunately, most of the public has little sense of what the risk factors really are. The chance of getting hit by lightning in the United States is about 5×10^{-7}, or 1 in 2,000,000, while the chance of getting killed in an automobile is much higher, 2×10^{-4}, or 1 in 5000. Yet people are far more concerned about lightning storms than getting into their automobiles.

Engineers know that things they design can hurt people. Automotive engineers recognize that over 44,000 Americans lose their lives every year due to accidents on the highways. Are the engineers then morally culpable for designing an unsafe product?

The key principle is that the level of safely must be understood and fully communicated to the user, and that any deviance from this accepted level of safety without the full understanding of the user is unethical conduct. The classical case of the Ford Pinto is illustrative. During the design of this car, the decision was made to put the gas tank under the trunk, behind the rear axle, even though this was a vulnerable position in such a small and "soft" car. The design met all the federal standards of the time, and the engineers did not intentionally build an unsafe car. On the other hand, they had built another car, the Capri, with the tank placed over the axle, so they had to have made a conscious decision to reduce the level of safety in the Pinto. When the rate of rear-end collisions resulting

(continued)

> **Box 14-1** *(continued)*
>
>
>
> in burn death climbed, they could have had a recall of all Pintos and an $11 strap on the tank would have solved the problem. But they chose not to recall the Pinto, calculating that it would cost less money for Ford to defend itself against all the lawsuits resulting from burn deaths than to fix the tank in a recall.
>
> In the Pinto case, the Ford engineers could be judged to have acted unethically because they produced a product that was significantly less safe than what the public expected, and when confronted with the facts, they chose not to fix the problem.
>
> More recently, the Firestone Tire Company produced tires that seemed to have a high rate of blowouts, especially in Ford Explorers. The Firestone Company, after initially denying responsibility, agreed to replace the questionable tires, not because the tires could not continue to be used on other cars, but because they believed (correctly) that the public would otherwise believe that all Firestone tires were unsafe. Ford Motor Company, however, continued to deny that the problem had anything to do with the Ford Explorer, even thought the statistical information was clear.
>
> When then is an automobile tire safe? It is not "safe" because it will never blow out and cause damage or even death, but it is "safe" when the incidence of such blowouts are no more than for comparable tires, and when the public *perceives* that this is the case.
>
> In our story, is it safe to walk down the streets of Atlanta? That's a tough call. It is perceived as safe as long as the rate of dangerous incidents is no more than in other cities. What you think is safe is based on your perceptions. When Great Aunt Meg told Chris, "Y'all just look safe, you *are* safe," however, she meant "Walk tall!"

You did not even own a car then, so she lent you her 1964 Galaxie. You had decided to visit Underground Atlanta, a supposedly revitalized downtown area in the only area of the city that General Sherman's troops didn't lay waste in their march through Georgia in November 1864. Driving past the Omni complex, you had noticed a 200-foot-wide billboard announcing that the Atlanta Police Department was understaffed, underpaid, and underfunded, and that you should take great care. In case anyone wondered why, the previous and (much higher) current year's statistics for murder, rape, and other major crimes were listed.

Nevertheless, you proceeded to Underground Atlanta, parking the Galaxie in an empty lot. You walked past what looked like an abandoned railroad station, exclusively populated by drunks, and hurried into what a guidebook assured you is a "vibrant urban renewal experience." Everything was closed down except for a handmade candle store run by a pair of Turkish brothers, who invited you to drink retsina and Turkish coffee with them, and informed you that they are the only business remaining. This was their last day in business, and when you got up to leave they insisted on giving you a beautiful candle![1]

It's all very different now, you think, as you wander around the fashionable boutiques, ethnic craft centers, and cafés that constitute today's revitalized inner city. Though you'd prefer not to have all those people of both genders asking if "You lookin' for a good time today, honey?"

Saturday, July 29

Gina has invited you to lunch on Saturday and, naturally, you spend much of it discussing Friday night's ball game. The score was tied 0–0 at the bottom of the eighth with John Smoltz, pitching a no-hitter, facing Mike Piazza. The Mets catcher was 0 and 2 on an inside

---

[1] Apart from fictitious Great Aunt Meg and the '64 Galaxie—it was a '72 Impala—this is based on ethicist Alastair's 1981 travel journal. The candle burned down long ago, but the taste of the retsina lingers on.

sinker and a super deceptive change-up when, inexplicably, Smoltz threw an 80-mph pitch right over the middle of the plate. Piazza, perhaps surprised, tried to crash it high over center field, but he couldn't have hit it quite right because Andruw Jones jumped high in the air and caught the ball just as it was about to sail across the fence. To celebrate, he hit a three-run homer in the ninth, and the Mets were swept for the second time in the season.

Over coffee, though, Gina becomes serious.

"Can we talk about a problem, please Chris?"

"Sure, what kind of problem?"

"It's about Sherm... no, not that kind of problem," she adds hastily. "Our marriage is fine. It's a work thing. You haven't met him, have you? So that gives you like an impartial view."

"Shoot."

"Sherm's wearing a lot of hats at the moment. You know he has an undergraduate degree in biology? After undergraduate school he bummed around for a few years, painted a bit, had a sign-writing business, did custom car painting, then he decided to get a serious qualification and got a master's degree in environmental engineering. Two years ago he passed his PE exam and became licensed. Took him ten years, on account of not having the undergraduate degree, and he was somewhat bitter about that."

Box 14-2

Professional Registration

In the days of the guilds (1500s to 1600s), one became a professional by surviving a long apprenticeship. The master provided all the schooling and eventually decided when one was ready to go off alone as a craftsman. The guilds covered almost all of the trades, from carpenters to stone masons and even fishermen. Modern engineering emerged with the establishment of technical schools, the Ecole Polytechnique in Paris being the first. Eventually, the minimum requirements for becoming an engineer included graduating from an engineering school, and this then became a necessary condition for professional registration.

Usually, right before graduation students are asked to take the Fundamentals of Engineering (FE) exam, a full-day written exam that covers most of the engineering sciences such as thermodynamics, chemistry, solid mechanics, electrical circuits, computers, as well as professional ethics.

Upon passing this exam, engineering graduates take on the role of apprentices and work alongside practicing engineers to learn the trade. States typically require an apprenticeship of four years in "responsible engineering work." In other words, the engineer has to show that this work was indeed in some area of engineering, using engineering skills and principles. At the conclusion of the apprenticeship, the engineer is allowed to take the Professional Engineering (PE) examination. This is also a full-day exam, but it is quite different from the FE exam. While the FE exam is mostly multiple-choice questions with short calculations, the PE exam also includes actual design problems requiring engineering solutions. In some cases, the questions include engineering ethics. Following the written exam, there is an interview with the candidates during which the general ethical and legal obligations of the profession are discussed. The PE exams given in the Canadian provinces, by contrast, are almost all on ethics, assuming that the applicants are technically competent to do engineering.

Upon successful completion of the PE exam, the engineer is licensed in the state in which the exam is given. Most states have reciprocal arrangements with others, so if one passes an exam in North Carolina, for example, one can simply apply for a professional engineering license in Pennsylvania. The only exception is California, which requires substantial proficiency in earthquake engineering.

The PE license has to be renewed annually, and in most states this requires proof of continuing education (attending workshops, short courses, professional meetings, etc.). With such continuing education, the PE license is granted for life.

(continued)

> **Box 14-2 (continued)**
>
>
>
> For people not having an undergraduate engineering degree from an accredited university, there is an alternative path to obtaining a license, but this requires a longer time in the apprenticeship grade. On applying to take the Fundamentals of Engineering exam, the applicant has to prove that he or she has been engaged in responsible engineering work, under supervision, for four or more years. Master's degrees often are allowed to count for two years' experience, but they do not substitute for the undergraduate degree because it is only the undergraduate degree that is accredited by ABET, the accrediting agency, not the master's degree. When the required years of work have been completed, the applicant is then allowed to take the FE exam, and upon passing this exam, the clock starts for the PE exam. This route often takes as long as ten years and substantial sacrifice in time and energy. In our story, this was Sherm's path to the professional engineering license.

"The Board does put a lot of weight on the degree," you reflect, thinking of some of your classmates who got through school mainly by copying other students' homeworks and labs.

"But Sherm has been lucky really. He has a lot of useful contacts, but I have to say he's a real quick study too. He started off doing contract work for the state EPA—one of his instructors at Georgia Tech works there now. The agency's understaffed and underfunded, at least that's what they keep telling everyone. And there was this sweeping new environmental law that required a lot of new policies. Sherm's master's dissertation was on the environmental engineering implications of the California law that our law was based on, this former professor was his supervisor, so he hired Sherm to write policy for them. Then someone at the Coastal Commission heard about him and put him on a retainer to have input on advisory rulings. He's working two–three days a week for the Commission."

"So he is formally a consultant to the Coastal Commission?" you say, beginning to see the outlines of the problem.

"Yes. But that's not enough to keep him busy. My company was approached to do some work for a client who wants to do some development on the coast. It's too small for us so we suggested a firm that we owed a favor to. Good company too, I might add. They were keen to do it, but they needed a specialist in a particular aspect of the new legislation, at short notice—and that's how Sherm started doing consulting work for this little firm."

You now see the problem clearly, but Gina continues.

"Here's the bad news. Last month Sherm got a letter from the State of Georgia Board of Professional Registration asking him to respond to a complaint, apparently sent in by some engineers, accusing Sherm of a conflict of interest."

> **Box 14-3**
>
>
>
> # Conflict of Interest I
>
> The literature on professional conflict of interest focuses on the difficulty of being "the servant of two masters," to be equally loyal to two parties. A conflict of interest may occur when, for example, someone
>
> - Works for or has a business interest in two organizations that are in competition—for instance, if one is an employee, a director, or consultant to two or more companies that are in the same business.
>
> *(continued)*

Box 14-3 (continued)

- Has a political position that enables the person to make decisions that might affect his or her business interests.
- Is an advocate or advisor for individuals or organizations who are applicants to an authority and is also employed by that authority.

The concept of a "Chinese wall" is sometimes used to refer to the mental processes whereby an individual is able to separate the potentially conflicting objectives of two organizations for whom he or she works. A professional company director who specializes in, say, the energy industry may well be a member of the board of directors of several competing energy companies. The idea is that when the director is acting for company A, he or she does not take into account the effects of the decision on the interests of company B of which she or he is also a director, or use "inside information" about company B when acting for company A.

However, not everyone is enamoured of this concept. Chinese walls are invisible except to those in whose minds they exist. They do not provide a transparent public reassurance.

Most codes of ethics of companies, institutions, and professional societies require employees and members to avoid conflict of interest, though not all of them define it in a helpful way. Some go into considerable detail, and the emphasis is often to ensure that a professional does not just act ethically but is *seen* to act ethically, avoiding any suggestion in the public mind of impropriety. As you read on, you may wonder just how far an organization may reasonably go to avoid the appearance of a conflict of interst. A good example is the University of Virginia gifts policy discussed earlier.

The University of North Carolina document *Responsible Conduct of Research* includes a Policy on Conflicts of Interest and Commitment. It states:

> The term conflict of interest refers to situations in which financial or other personal considerations may compromise, or appear to compromise, a faculty member's ... professional judgement in exercising any University duty or responsibility or in conducting or reporting research.

Stanford University's Faculty Policy on Conflict of Commitment and Interest states:

> A conflict of interest occurs when there is a divergence between an individual's private interests and his or her professional obligations to the University such that an independent observer might reasonably question whether the individual's professional actions or decisions are determined by considerations of personal gain, financial or otherwise. A conflict of interest depends on the situation, and not on the character or actions of the individual.

"Do *you* think he's in a conflict of interest? Presumably he makes sure he doesn't work for, say, a developer and the Coastal Commission on the same project?" you ask.

"It's not that simple. Take policy advice, it has an impact on everything. And how can he oppose coastal development and work for a client who wants to build a marina? Funny thing is, his client is totally unfazed. When he works for the client he gives it a hundred percent, and they know it. And they know that Sherm is very knowledgeable about the new regulations so he can keep the developer out of unforeseen regulatory problems. He also thinks it makes them look more, like, environmentally virtuous, having a committee member of SOB working for them, and ..."

"SOB?"

"Save Our Beaches, yeah, a community action group opposed to further development on the beaches. SOB thinks that having Sherm on their board makes SOB look good, more moderate, and the developer thinks that having Sherm as his consultant makes him look good. I'm not sure if the Coastal Commission knows that Sherm is also working for the developer, and I'm not sure they would care. I guess everyone thinks he'll come across with inside information." (See Box 13-1.)

"And does he? How does Sherm feel, wearing all those hats?"

"No, he certainly doesn't, and he feels fine about it all. He says he's doing freelance engineering, no conflict of interest at all, he gives impartial, professional advice to anyone who wants it—at a price to his employers and clients, and pro bono to SOB. He says it's all a matter of Chinese walls in your head."

"He's talked to an attorney?"

"Course. Hartmann B. Duffelberger of Robinson, Tollemache, Gobsmacker."

"Really?"

"Along those lines. Two hundred an hour. No idea what he *really* thinks, that's not what you pay lawyers for, but the lawyer has written letters to the State Board of Professional Registration demanding they withdraw the charges."

"Get anywhere?"

"The Board? They're stalling at the moment, but a friend told Sherm she'd heard they were taking advice from an ethics consultant, whatever that is. They probably charge two hundred an hour too. I didn't think there were any such things as ethics experts. I mean it's not like structural design where you've got right and wrong answers." (See Box 4-1.)

This sounds familiar.

"What does Sherm really want?" you ask.

"He wants to go on with all of his present activities, so long as his clients are happy. He thinks he has the right to work for whoever he likes, and he enjoys the variety. But he also wants to keep his PE license. And, I suspect, he wants the ethical approval of his peers. Still, I guess he can't have it all ways."

That's familiar too.

"So," Gina says. "What do you think?"

You think for a moment.

"It sounds complicated. I haven't read the State of Georgia professional engineering Code of Ethics, so I wouldn't know what the exact wording is, but if it is like most engineering codes, it looks like Sherm has little choice. Either he changes his professional practice to conform with the code, or he loses his PE license. If he loses his license, he can still work as a consultant as long as he does not pass himself off as a professional engineer."

Box 14-4

Conflict of Interest II

All engineering codes of ethics have statements on conflicts of interest. The ASCE Code of Ethics, for example, deals with this topic under canon 4:

CANON 4.

Engineers shall act in professional matters for each employer or client as faithful agents or trustees, and shall avoid conflicts of interest.

a. Engineers shall avoid all known or potential conflicts of interest with their employers or clients and shall promptly inform their employers or clients of any business association, interests, or circumstances which could influence their judgment or the quality of their services.

b. Engineers shall not accept compensation from more than one party for services on the same project, or for services pertaining to the same project, unless the circumstances are fully disclosed to and agreed to, by all interested parties. . . .

d. Engineers in public service as members, advisors, or employees of a governmental body or department shall not participate in considerations or actions with respect to services solicited or provided by them or their organization in private or public engineering practice. . . .

(continued)

Box 14-4 (continued)

g. Engineers shall not accept professional employment outside of their regular work or interest without the knowledge of their employers.

Like the rest of the Code, this is rather general and, importantly, does not actually define "conflict of interest."

The National Institute for Engineering Ethics (NIEE) states:

> Engineers shall act as faithful agents for their employers or clients and maintain confidentiality; they shall avoid conflicts of interest whenever possible, disclosing unavoidable conflicts.

Once again, there is no definition. However, NIEE makes available a range of useful material on its web site, www.niee.org, including a case that we reproduce below. It is an interesting example of how a Board of Ethical Review goes about applying a provision in a code of ethics to a set of facts. (www.niee.org/cases/78–88/Case%2079–1.htm)

> This opinion is based on data submitted to the Board of Ethical Review and does not necessarily represent all of the pertinent facts when applied to a specific case. This opinion is for educational purposes only and should not be construed as expressing any opinion on the ethics of specific individuals. This opinion may be reprinted without further permission, provided that this statement is included before or after the text of the case.

Conflict of Interest—Payment from Related Party   BER Case 79-1

Facts:

A government agency retained an architectural-engineering firm as the prime professional for the design of a major hospital. The A/E[2] firm, in turn, retained Engineer A as a structural engineer consultant and Engineer B as a concrete consultant. The agreement with the A/E firm required, among other things, that the A/E firm review and approve shop drawings furnished either by the government or the contractors for conformity with the design concept and the contract documents. This duty was contracted by the A/E to Engineers A and B for their respective portions of the work. The steel supplier then retained Engineer A to prepare the shop drawings for its portion of the work. The concrete subcontractor retained Engineer B to prepare the shop drawings for that part of the work. Engineer B also agreed to serve as an expert witness on behalf of the concrete subcontractor in the event of any dispute for that part of the work. The responsible engineer of the A/E firm was advised of these arrangements and made no objection. The government agency learned of the facts before construction started and has questioned the ethical propriety of these arrangements.

Questions:

Was it ethically permissible for Engineers A and B to enter into the above arrangements with the construction subcontractors?

Was it ethically permissible for the responsible engineer of the A/E firm to permit these arrangements?

References:

Code of Ethics—Section 8—"The Engineer will endeavor to avoid a conflict of interest with his employer or client, but when unavoidable, the Engineer shall fully disclose the circumstances to his employer or client."

Section 8(a)—"The Engineer will inform his client or employer of any business connections, interests, or circumstances which may be deemed as influencing his judgment or the quality of his services to his client or employer."

Section 8(b)—"When in public service as a member, advisor, or employee of a governmental body or department, an Engineer shall not participate in considerations or actions with respect to services provided by him or his organization in private engineering practice."

Section 10—"The Engineer will not accept compensation, financial or otherwise, from more than one interested party for the same service, or for services pertaining to the same work, unless there is full disclosure to and consent of all interested parties."

Section 10(b)—"He will not accept commissions or allowances, directly or indirectly, from contractors or other parties dealing with his clients or employer in connection with work for which he is responsible."

Discussion:

In a related case, 72-9, we noted that an engineer's dual role as the owner's agent, through an architect, and a bidder's agent, through a subcontractor, created an obvious conflict of interest. We added that it is not unethical per se to become involved in a conflict of interest under §8 of the Code, but that it may be unethical depending on the circumstances. "The avoidability of a conflict of interest is a subjective judgment. Its impact on the consideration of a case involving §8 will vary according to the circumstances." In that case, arising under quite different facts, we concluded that the engineer had not acted unethically

(continued)

---

[2]A/E is Architectural/Engineering

Box 14-4 (continued)

because of the time elements involved which, in effect, made the conflict unavoidable. There is no similar time element in the case before us, as indicated by the question raised by the governmental agency representative before construction work started, hence the conflict was avoidable.

The purpose of the shop drawing clause, which is commonly used in engineering documents, is to protect the interests of the owner by assuring independent review of the shop drawings as part of the duty of the design professional to determine on a timely basis that the project will be built to comport with the design concept. Obviously Engineers A and B could not "independently" review the shop drawings they had prepared. It should also be noted that the duty to review the shop drawings was imposed on the A/E prime professional, and while this function could be delegated to associate engineers serving as consultants that delegation did not change either the legal or ethical duty of the prime professional.

It is clear that Engineers A and B are in conflict with §10(b) under their respective arrangements to receive compensation from both the owner (through the A/E) and the steel supplier and concrete subcontractor for services pertaining to the same work. As we are advised, the A/E prime professional has been made aware of the arrangements by Engineers A and B, and has consented by not objecting. But that is not a sufficient answer to the requirement of §10 that there be both full disclosure and consent of all "interested parties." The owner is clearly an "interested party" and by raising the ethical issue has clearly not consented. It was incumbent upon Engineers A and B to assure themselves that their arrangements were approved by the owner as an "interested party."

Aside from the disclosure requirements of §8, §8(a) imposes the further obligation on the engineer to inform his client of any business relationships which may be considered as possibly influencing his professional judgment. This has been done by Engineers A and B, although it is not clear that they would agree that their respective relationships might influence their judgment or the quality of their services. We have also cited §8(b) of the code, which applies only in principle, because Engineers A and B are not in public service in the intent of that provision. It is pertinent, however, as being consistent with the philosophy of the other portions of §8 to indicate the importance of engineers not being in a position of trying to serve opposing interests.

The agreement of Engineer B to also serve as an expert witness on behalf of the concrete subcontractor in the event of a dispute on that part of the work is so patently in conflict with ethical precepts that it does not warrant extended discussion. It should have been apparent to Engineer B that he cannot appear on behalf of the subcontractor while performing his primary ethical duty to represent the interests of his client, whether that "client" be interpreted only as the A/E prime professional or as the ultimate client, i.e., the government agency.

Under these circumstances and considerations, we believe that Engineers A and B will compromise, or appear to compromise, their independent position and professional judgment by attempting to serve the primary interests of the owner and the steel supplier and concrete subcontractor at the same time on the same project.

Conclusions:

It was not ethically permissible for Engineers A and B to enter into the above described arrangements. It was not ethically permissible for the responsible engineer of the A/E firm to permit these arrangements.

You think some more.

"Maybe he could wait for the Board to contact him again. If they're really getting an ethics opinion, presumably they'll communicate it to him and advise him of their thinking. They won't want to rush things, and they'll surely give him plenty of opportunity to defend himself."

"His attorney said that, too," Gina replies. "He told him there have been successful lawsuits against professional boards of registration that have denied people membership without good cause, on account of it damages people's reputation and makes it harder for them to get work. Usually settled out of court, of course, once the billable hours have reached an agreed level."

"His attorney said that?"

"Naw, I just get a charge out of insulting lawyers. How come they always get shown as crusading heroes in movies and on TV, Perry Mason, Susan Sarandon defending little kids for free, Tom Cruise doping out his crooked boss?"

"He's just about the only honest person in the firm."

"OK, but you're supposed to go away thinking he represents all that's most noble about the profession."

"Maybe lawyers are like engineers—you only hear about them when they screw up, or wimp out on public safety issues," a thought that comes much too close to your own concerns.

"Anyway . . ." you continue. "It seems to me that Sherm needn't decide anything right now. I would suggest, however, that he tone down the environmental activism. If it comes to a serious confrontation he'll have no real choice anyway, and by then he may have lost credibility with both sides."

"*All* sides!" corrects Gina. "And thanks for the advice," she continues. "Sincerely, you've been a big help. I knew I could count on you for good advice on ethics. Your career seems to be on the right course, and I have always thought of you as the best in professional engineering. It's good to have heroes!"

On course for what, you wonder. But it feels good to be thought of in that way and you hope you are not blushing.

Sunday, July 30

You, Gina, and Sherman watch Atlanta win easily against Cincinnati. Ryan Klesko and Chipper Jones hit back-to-back homers in the fourth, Andruw Jones hits for the cycle including a stand-up triple with two RBIs in the sixth, and Rich Gossage, whom the Braves have lured out of retirement because of injuries to all their closers, strikes out the last five Cincinnati batters. Marje Schott is ejected in the middle of the eighth for arguing a pitch with the home plate umpire, whom she calls a Nazi. (Next day the *Atlanta Constitution-Journal* runs the story under the headline "Goose steps out in style: Marje shot.")

During an inning break, while Gina goes to get something to drink, Sherman says quietly to you, "Gina told me about your discussion yesterday."

Sherman is a surprise. After what Gina told you about him, you expected a weedy, hippy-looking guy with a ponytail, but in fact he is wearing (to a baseball game!) a blue Armani suit, a white shirt, and exquisite blue and white silk tie and Italian shoes that must have cost him at least $300.

"I hope you didn't mind us talking about you," you say.

"No, no, I need all the advice I can get," says Sherman. "And I'd like to thank you. But I disagree with your suggestion that I can't work for both the developer and the Coastal Commission. I don't see why I should turn down a job just because I'm also helping someone else. I can do a good job for both, and in this case I am clearly the best person to do both jobs."

"Maybe," you say. "But the problem is with perception. This is why the engineering codes forbid conflicts of interest." You give him a serious look. "It is the *perception* of the conflict, not the actual action that is important."

"That's just bullshit!" Sherman retorts, getting louder and more animated. "It took me a long time to get to where I am making some serious money in engineering, and I am not about to give this up just because there is a *perception* of impropriety."

You don't know what to say. He is, after all, your host and the husband of a professional friend. But he clearly does not understand the concept of professional engineering. You hear him go on:

"Listen, I worked hard to get my license. Ten years it took me because of the lack of an undergraduate degree. I was doing more and better engineering than many of my coworkers, but they had the undergraduate degree, so they become PEs after four years. I had to wait ten years and then convince the old farts on the Board that I was worthy of a license. And now I am going to cash it in. I'm sorry I don't have the money or the time to be what you call 'ethical.'"

Box 14-5

Why Be a Good Engineer?

Engineering, because it always involves people, is more than just the application of technology. Engineers who can perform their assigned technical tasks are competent and capable. But we would like to argue that, in addition to possessing technical skill, a *good* engineer is well mannered, has high moral standards, and obeys the law.

As we argue throughout this book, everyone, given the choice, wants to live in a society where everyone agrees to abide by good manners, high morals, and lawfulness. If everyone buys into this agreement, then everyone benefits. Those societies where human dignity and reciprocity in manners, morals, and laws was not practiced—for example, the Hutu massacring Tutsi and each other, the Serbs achieving ethnic purity by killing the Albanians, the Nazis ridding themselves of non-Aryans, death squads killing off thousands of young people in Chile, Stalin murdering 30 million of his own people in death camps—are not societies where anyone wants to live. There is a great deal to be gained by practicing and promoting civility.

This is also true in professional engineering. In order to help maintain a viable engineering profession, engineers should demonstrate good professional manners and, if the occasion requires, admonish others for boorish behavior. They should act as role models in conducting engineering on a high moral level and promote such behavior in others. And without doubt, they should not become criminals. In short, engineers all have a responsibility to uphold the honor of professional engineering—to create a culture in which all engineers can flourish.

But there is a larger question of why any engineers should *act* in such an honorable way. That is, if engineers agree that having bad manners, or acting immorally, or even breaking a law is advantageous individually, why should a specific engineer, at any given moment, *act* honorably? That is, why practice good manners, act morally, or obey the law when the opportunity occurs to do otherwise, especially when dishonorable behavior would seem to be personally beneficial?

The answer comes in several parts. First, nobody wants to get caught and suffer the consequences. Bad manners would result in ridicule; immoral conduct might result in loss of clients and business; and of course breaking a law might result in a fine or jail time.

Second, engineers recognize that they are all members of a larger community, in this case the engineering community, and they all benefit from this association. Acting in a manner that brings harm or discredit to this community cannot, in the long run, be beneficial. Granted, the destruction of professional engineering may be far into the future and small antisocial acts would not be enough to destroy the profession, but all engineers nevertheless have obligations to uphold the integrity of the profession. In short, engineers should act honorably because the profession depends on them to do so.

Third, any antisocial act always comes at a personal cost. Sissela Bok (1978), in her book on lying, contemplates the decision to lie or not to lie in a circumstance in which the lie will result in apparent greater good at little cost to the teller. For example, she agrees that it is obligatory to tell a lie to save an innocent life, but points out that every time a lie is told the teller is less of an honorable human being. There is, as it were, a reservoir of good in each human, and this can be nibbled away one justified lie at a time until the person is incapable of differentiating between lying and being truthful. This would also be true for manners and legal acts. Every lie (or any other antisocial act) reduces one's own standing as an honorable human being.

(continued)

Box 14-4 (continued)

Finally, the reason for not being antisocial, even assuming we could get away with it, is that eventually the human conscience would not stand for it. Michael Pritchard (1991) argues persuasively that everyone has a conscience and is able to differentiate between right and wrong. Although most people occasionally act dishonorably, they *know* they are behaving badly and eventually regret such actions. Thus, dishonorable actions will always result in personal harm. Similarly, engineers who behave without regard to manners, morals, or laws will eventually cause harm to befall upon themselves. They *will* lose clients, and their untruths *will* cause their works to fail, and they *will* have a bad conscience that will bother them. They will eventually think poorly of their own standing in the profession and regret their selfish, self-serving actions that may have been ill mannered, immoral, or illegal.

But what if, in the face of these arguments, one is still not convinced that acting honorably is a good thing to do? Sadly, there appears to be no knock-down ethical argument available to change the mind of a person set on behaving badly. Humans have the option to act in any way they wish, even if the action is evidently self-destructive.

In our story, Chris is being asked to think hard about ethics and professionalism. Most important, Chris's opinion seems to be valued by Gina. In this case, Chris's advice is good advice, both in terms of the moral standards most people would accept and also in terms of the engineering code of ethics. In the future, if Gina were ever to reflect on how this issue played out, she would think of Chris as a very good person who gave her good advice, and Chris is no doubt glad to be thought of as a good person.

Wishing to be thought of as a good person is not a new idea. For example, while the Viking society of northern Europe was in many ways cruel and crude by our standards, they had a very simple code of honor. Their goal was to live their life so that when they died, others would say "He was a good man." Their definition of a "good man" might be quite different by contemporary standards, but the principle is important. To professional engineers who conduct themselves so as to uphold the exemplary values of engineering, the greatest professional honor imaginable would be to be remembered as a *good* engineer.

Discussion Questions

14-1. Consider Chris's state of mind at the end of this chapter. What was Chris thinking? What effect do you think Gina's request for help had on Chris's self-evaluation?

14-2. Describe some actions that are (a) bad manners, but not immoral or illegal; (b) immoral, but not bad manners or illegal; and (c) illegal, but not immoral or bad manners.

14-3. Codes of ethics are discussed in several places in this book. Some people claim that such codes are no more than pieces of paper—irrelevant and unenforceable. Do you agree? How could a code of ethics be made more useful, if this is possible?

14-4. Do you agree with Sherman? Why cannot he continue to advise the various groups—the developer, the regulatory commission, and the environmental group—as long as he sincerely knows that he is giving them all his best advice?

14-5. The Board of Professional Registration sent Sherman a letter telling him that there was a complaint lodged against him for alleged conflict of interest. The letter described such a situation in detail and indicated why this is a conflict of interest. Write such a letter, from the State of Georgia Board of Professional Registration, to Sherman Robbins, making sure you clearly outline the problem.

14-6. As we note in several places in this book, engineer Aarne and ethicist Alastair have different views about the nature of ethics: Briefly, Aarne believes in universal ethics, while Alastair is a cultural pluralist. Box 14-4 is a compromise of these views. What is your reaction to this? How can we defend what many people regard as universal standards while making appropriate allowances for cultural differences?

References

Bok, S. 1978. *Lying: Moral Choice in Public and Private Life*. New York: Pantheon Books.

Pritchard, M. S. 1991. *On Being Responsible*. Lawrence: University Press of Kansas.

15

Objective and truthful manner

Sunday, September 15

You're sitting in your yard, when the phone rings. Alex answers it, and there's a silence followed by a loud "Oh, no!" You rush inside and Alex hands you the phone. It's Sarah. "I don't know how to tell you this," she says, "but, well, you were right about the overhang, the Asmara. There's been a structural failure, and . . ."

"How bad is it?" you ask.

"It's on CNN right now."

Shocked, you turn on the TV. A reporter is standing in front of a pile of rubble, from which rescue workers are pulling bodies. Police, paramedics, and construction workers are rushing around as the reporter says, " . . . four definite fatalities so far, Mike, up to 20 hurt, and an unknown number trapped in the rubble. So far there's no evidence of a bomb or other terrorist involvement, and at this stage, Mike, it looks as though the tragedy may have been caused by a construction failure. The governor arrived about ten minutes ago, and we hope to bring you an interview with him momentarily."

The camera switches to the studio where an anchor is summarizing the tragedy, concluding with ". . . and now back to Moyna at the scene of the Hotel Asmara disaster in Philadelphia."

"Thanks, Mike. I'm talking to state governor Tom DiStefano. Governor DiStefano, how did you feel when you first heard the news of this disaster, sir?" The harried-looking DiStefano, wearing a business suit and a hard hat, mouths phrases like "terrible disaster," "deepest sympathy," and the like. Then his face hardens and he says crisply, "Whether this is the result of a terrorist act or a construction failure, it's a tragedy and a crime. And whoever is responsible will be found and made to pay to the utmost severity of the law."

"Thanks for your time, Governor. Mike, the Asmara opened only last year, and this is the first hint of any problems. The design company, Pines, and Timmo, the builder, have excellent reputations and they've never had another building collapse on them."

"Thanks, Moyna. Of course we can't pass judgement until we know the cause of this terrible disaster." Sounds like you already did, you think bitterly. "We're trying to get a comment from Pines, the company that designed the hotel, but at this time all of their senior people are unavailable to speak with us. However, we do have here an expert from design firm Randle Associates who has agreed to speak with us."

A nervous-looking young man—he looks only a year or so out of engineering school—appears on the screen, accompanied by a caption reading "Matthew Higgins, structural engineer."

"Mr. Higgins, thanks for agreeing to give us your views. How did you feel when you heard about this disaster?"

Clearly uncomfortable at having to speak on camera, Higgins says, "Well, naturally I was shocked. It's a terrible tragedy."

"Do you have any views about the reason, why it occurred?"

"Um, I couldn't say, that will be for an inquiry to determine."

"Yes, of course, but in your opinion, as a structural engineer, what would have caused the collapse of this overhang?"

"Well, things don't just collapse on their own. There must have been a mistake somewhere."

"You mean the design was wrong?"

"If the contractor is not at fault, ah . . . , well, then the design must have been wrong."

"Pines Engineering did the design. What is your opinion of this firm?"

"OK, I guess."

"Just OK? That does not speak very highly of the firm."

"Well, I have heard some foremen on another job talking about how Pines has screwed up the design at other construction jobs. Maybe they messed up here also."

Box 15-1

Professional Respect

The first professional engineering societies in the United States were founded in Boston and New York. Following a period of bickering and one-upmanship, they finally merged in 1852 into one society named the American Society of Civil Engineers (ASCE). At the time of ASCE's founding, there was some informal discussion of developing a code of ethics, but the society decided not to adopt such a code and to rely on the personal dignity and honor of the individual engineer. After many false starts, the first ASCE Code of Ethics was adopted in 1914.

CODE OF ETHICS

Adopted by the Society by letter ballot,

September 20, 1914

It shall be considered unprofessional and inconsistent with honorable and dignified bearing for any member of the American Society of Civil Engineers:

1. To act for his clients in professional matters otherwise than as a faithful agent or trustee, or to accept any remuneration other than his stated charges for services rendered his clients.

2. To attempt to injure falsely or maliciously, directly or indirectly, the professional reputation, prospects, or business of another Engineer.

3. To attempt to supplant another Engineer after definite steps have been taken toward his employment.

4. To compete with another Engineer for employment on the basis of professional charges, by reducing his usual charges and in this manner attempting to underbid after being informed of the charges named by another.

5. To review the work of another Engineer for the same client, except with the knowledge or consent of such Engineer, or unless the connection of such Engineer with the work has been terminated.

6. To advertise in self-laudatory language, or in any other manner derogatory to the dignity of the Profession.

This code is directed solely to interactions between engineers and does not in any of the clauses recognize that the engineer is responsible to society. Note the second item. Even in 1914, it was considered unethical to speak badly of fellow engineers.

Engineering *is* a profession. As such, society yields certain privileges to the engineer and expects in return that the profession will serve the best interests of society. Like the physician keeping his or her clients

(continued)

Box 15-1 (continued)

healthy, and the lawyer keeping his or her clients out of jail, the engineer is expected to build a physical societal infrastructure in which his or her clients can live and prosper. All professions recognize the need for society to respect and believe in its individual practitioners. To speak ill of fellow professionals is therefore frowned upon, since it diminishes the luster of the profession and can lead to lower pay and status, and less freedom to run its own profession.

In order to underscore this point, the ASCE Code of Ethics was modified in 1942 to include the eighth article, which stated that it was considered unprofessional

> 8. To act in any manner or engage in any practice which will tend to bring discredit on the honor of the Engineering Profession.

In a way, this is a curious change since it does not seem to add anything: It can be boiled down to the tautologous admonition "it is dishonorable to be dishonorable," or "it is not dignified to bring discredit on the dignity of the profession." Nevertheless, the important point is that this article recognized that there is a public out there that might care about how engineers conduct themselves.

Why this preoccupation with the dignity of the profession and the manner in which engineers treat each other? Simply put, professional engineers are granted special privilege by society (such as being gatekeepers to their own profession), and they like it the way it is. In addition, engineers argue that a professional of high standing will draw better people to the profession and hence result in a higher level of service to the public. Speaking ill of one's fellow engineer is therefore severely frowned upon.

The camera cuts to the reporter who is summing up the interview.

"While the cause of the collapse is still undetermined, it seems that the engineers for this building have a history of shoddy engineering design, and every indication is that they are at fault in this instance as well. Back to you, Mike."

Box 15-2

Engineers and the Media

Journalists walk a narrow line between professionalism and sensationalism. Most journalists are professionals, seeing themselves as a link between the public and the world, including government, science, entertainment, and society in general. They vehemently and eloquently defend the journalistic enterprise as a high calling, are careful and fair in what they print or what they say, and recognize the importance of their role in upholding our First Amendment rights. Times without number, journalists have exposed dishonesty, inefficiency, corruption, and disaster in business, local and national government, sports, the churches, international organizations, and so on.

But journalism also has a less professional side. When the pressures of deadlines and money dictate journalistic activities, even some of the most well regarded news organizations can stumble and place expediency ahead of professionalism. All news organizations (in a free society) recognize that if nobody bought the papers, and if nobody watched the tube, then there would be no newspapers or television.

Such pressures are a constant concern for even the most modest hometown papers and radio and television stations. Most of the reporters that engineers encounter work for the local press, and it is on the local level where engineers are most visible to the public and where engineering most often influences people's everyday lives. Unfortunately, engineers often do not understand that the role of a reporter is very different from that of a professional engineer and treat reporters as they would their professional engineering colleagues. Many engineers are unaware of the pressure on reporters to publish what editors demand or whatever sells more newspapers or commercials.

(continued)

> **Box 15-1** *(continued)*

Local reporters for both the print and electronic media work for a living, and they get paid for obtaining newsworthy material. While most reporters are interested in relating a balanced account of a news story, some will also try to get the principals (often an engineer) to say something that will result in headlines or sensational sound bites. At times, the engineer can use the reporter to further his or her own objectives (such as preparing for a public hearing), but other times, the reporter can get the engineer to say something that is not in the engineer's best interest. The reporter views this as simply doing a good job, and usually there is no malicious intent.

Reporters admit that during an interview, the interviewee (the engineer) is in the power position. The journalist will do everything possible to reverse those roles, and it is up to the engineer to prevent the switch.

Engineers should understand what the reporter's perspective is and respond accordingly. In general, the media distinguishes factual reporting from editorializing, but personal opinion can creep into ostensibly factual news stories. Every person, reporters included, has a perspective based on background or education, and this may be quite evident in the reporting. Even the decision to do a story is a value judgment because this defines what is and is not news.

Engineers should recognize that there are three modes of interviewing: on the record, off the record, and background. If the interviewee agrees to be *on the record*, then everything said can be used and attributed directly. If it is *off the record*, then nothing can be attributed directly ("A senior White House aide confirmed that . . . ,"). *Background* means that it is not permissible to even acknowledge that the conversation has taken place. The purpose of a background interview is to educate the reporters so they can better report the stories.

If a reporter breaks a promise of confidentiality (for example, attributes an interview that was to be "off the record"), he or she will be severely criticized by the journalistic profession, and it is unlikely that the reporter will ever again have the chance to interview the person involved. These agreements are necessary for both the press and the public, and breaking the rules will result in severe consequences.

Most important, engineers should not speak publicly unless they are knowledgeable. If an engineer is not the best person to answer a question, he or she should suggest that the reporter ask the person who is. In our story, Matthew Higgins was not in a position to know what happened to the hotel overhang, and he decided nevertheless to speculate on camera. He should have said, "I really cannot comment because I don't know the facts," if he was forced to be on camera, but the best action would have been to not agree to go on camera at all. His actions represented unethical professional conduct by any engineering code of ethics.

"Thanks, Moyna. We'll keep you posted on the Asmara tragedy, but now we move to the latest disaster in war-ravaged Central Africa, where the threat of famine is looming into a crisis of unknown proportions."

The telephone rings. You decide not to answer it, and you hear the voice mail greeting come on. There is a pause after the greeting, as if the caller is trying to find the right words.

"Chris, this is Shawn," the voice says. "You remember me? I'm the lawyer you consulted on the ethical problem you had. I think you might need my help. Why don't you stop by my office tomorrow morning?"

Discussion Questions

15-1. The ASCE Code of Ethics states that it is in effect unethical to speak badly of fellow engineers. Why do you suppose this is one of the main tenets of the Code of Ethics? Is this constraint self-serving, or is this admonition in the best interest of the public?

15-2. What do you think will happen to structural engineer Matthew Higgins? Imagine the next time he comes to the meeting of the local engineering society. How do you think he will be treated by the other engineers? Why? Would their actions be justified?

15-3. We have tried to describe the CNN report as it might have happened. Do you think we have presented it fairly? What issues of journalistic ethics does our account suggest?

Epilogue

We will not pursue this story any further, because its ending will be determined by the legal process and this is not a book about legal ethics—that's another book!

But in a nutshell, this is what will happen: The District Attorney's office will conduct an inquiry to determine whether there was any criminal negligence. In this case, there certainly was, and hearings will be held and charges will be laid. The engineers involved will be asked if they knew about the poor design, and they will have to either commit perjury or own up to the fact that they did know. Kelly will be called as a witness. Shawn will not because, as a lawyer, he cannot be asked to reveal details of discussions with his client, which are privileged. Chris will probably be held to be an accomplice but may through plea bargaining or turning informant be able to escape prosecution—unlike Joe, Sarah, and Ken. They face serious charges and may go to jail or at least have to pay heavy fines (which Pines' insurance company will pay). Pines will disappear. The engineers will all lose their PE licenses, with little hope of getting them back.

Simultaneously, the insurance companies involved will have their own investigations in order to assign blame (you may feel that a more important priority would be to figure out how to prevent similar disasters from occurring in the future, but this is not the system—blame first, then corrective action).

Once blame has been established by hearings and/or courts, the survivors of the people killed will start lawsuits. This will happen even in the unlikely event that no one is found to be criminally negligent because the standard of proof is lower in civil suits than in criminal cases—as happened in the case of O. J. Simpson, who was acquitted of murder but was required to pay civil damages to the families of the people whose death the jury found he had wrongfully caused. These suits will go on for years, exhausting all concerned emotionally—but not financially. The lawyers for the families will all be working on a contingency basis, where they are paid a percentage of the final settlement, and the engineers will all be covered by Pines' insurer. Eventually, the insurance company will probably settle out of court, and the families and their lawyers will receive millions of dollars. So it goes.

Where are they now?

- *Ah Chee* is working on a resort in Tierra del Fuego.
- *Arthur* (Kelly's boss) got into politics and became the Assistant Administrator for Water Programs at the U.S. EPA.
- *Alex* is still at McGregor University. The new president has asked Alex to be the associate dean.
- *Atlanta* becomes the first team ever to win the World Series four straight years.
- *Chris* rents a small house on the Outer Banks, North Carolina, and lives there year-round, drinking a bottle of Beam a day, and working on a book about engineering ethics. Local gossip has it that Chris and Kim (of Kim's Krabkakes) have a thing going.
- *Dave* becomes a Jehovah's Witness and lives in Cairns, Australia, where he is a sexual abuse counselor.
- *Earl* is run down by an ice cream truck at Myrtle Beach, South Carolina, and is in a persistent vegetative state. His elderly parents are bringing suit to have his life support disconnected.
- *Freddie* and *Ginge* sell Freddie's and move to Atlanta. Their Norm'n'Beas vegetarian fast-food franchise operation is very successful.
- *Gina* gives up on Sherman and becomes chief design engineer at P20's Colombian branch. She lives in Bogota with a distant cousin of Edgar Renteria Jr.
- *Girish* is appointed minister for the environment in Genala.
- *Ivan* retires from academia and becomes a partner in a firm of immigration consultants specializing in recruiting technical people and engineering students from Southeast Asia. De Tocqueville University is one of his biggest clients.
- *Joe* cooperates fully with the authorities and pleads nolo contendere to a minor charge of violating safety standards, for which he is fined $50,000. He successfully relaunches Pines as Dogwood Structures.
- *Ken* is sentenced to 18 months in a federal prison. He is released after serving 7 months, and he and his wife retire to Naples, Florida.
- *Kelly* returns to Southern California where she is a successful consulting environmental engineer and an adviser on professional ethics to ASCE.
- *Mario* returns to his wife and family in Manila.
- *Sarah* cooperates fully with the authorities and escapes criminal charges but loses her PE license. She retires from engineering practice and becomes a successful real estate agent, eventually endowing the Department of Civil Engineering at De Tocqueville University.
- *Sherman* gives up planning and goes to work full-time for Save Our Beaches. He is arrested four times for arson and criminal damage at coastal construction sites. Gina sometimes sends him money.
- *Trish* and *Ben* marry and move to Oshkosh, Wisconsin, where they have a successful construction business. They regularly holiday in their Genalan time-share.

- And you, the reader, still don't know Chris's or Alex's gender.

Index

AAUP, 78
ABA, 13
ABET, 90,115,140
Academic freedom, 78
Active euthanasia, 33
Active whistle-blowing, 62
Adams, Scott, 15, 80, 84
Admonitions, in codes of ethics, 15
Advertising, 89
Affirmative action, 116
AIChE, 13
AIME, 13
Alpern, Kenneth, 12, 63
Altruism, 20
Alumni whistle-blowing, 62
AMA, 13, 34
Animals, 56
Anonymous whistle-blowing, 62
Anthropocentrism, 57
Arafat, Yassar, 80
Architects, 12, 67
Aristotle, 21
Armaments, 107
Armstrong, Neil, 80
Arthur, Chester, 44
ASCE, 13, 16, 17, 92, 97, 128, 142, 150
   Code of Ethics, 93, 97, 128, 142, 150
ASME, 13
Assault rifle, 108
Australia, 34, 38, 56
Automobile fatalities, 137

B-2 bomber, 108
Background, 152
Beach Boys, 105
Beastie Boys, 105
Bentham, Jeremy, 20
Berkeley, University of California, 99
Bible, 4, 6
Bidding, 92
Biocentrism, 57
Blind loyalty, 134
Bok, Sissela, 96, 125, 146
Boysjoli, Roger, 46
Bowie, Norman, 96
Bresznev, Leonid, 80
Bribery, 95

British Commonwealth, 35
Britten, Benjamin, 105
Broome, Taft, 61
Brontland Commission, 17
Brooklyn Bridge, 13, 28
Brown and Williamson, 63
Brunel, Marc Isambard, 80
Bucknell University, 112
Buddah, 124
Buddhism, 7
Burma, 124

Camus, Albert, 18
California, 139
California, University of, 99
Canada, 79
Carter, Jimmy, 80
Categorical imperative, 20
Challenger, 46, 47
Chapel Hill NC, 121
Charity, 7
Charles, Baron de Montesquieu, 123
Cheating, 6
Chernobyl, 29
China, 124
Christianity, 4, 5, 123
Chinese wall, 141
Citicorp building, 45, 48
Civil engineering, origins of, 13, 28
Classical utilitarianism, 20,22
Clinton, William, 125
Coca-Cola, 112
Code of Ethics, 14, 16, 27, 93, 97, 128, 142, 150
Collingwood, Francis, 44
Competitive bidding, 92
Conflict of interest, 140, 142
Consequentialist ethics, 19
Cooper Union, 13
Corman, Roger, 80
Cost-benefit analysis, 60
Cultural pluralism, 97, 100

da Vinci, Leonardo, 28
Dead loads, 40
Death with Dignity Act, 34
Deception, 2, 71, 83
Decision making, 51, 52

DeGeorge, Richard, 62, 70, 79
de Montesquieu, Charles, Baron, 123
Deontological ethics, 20
deRobertis, 105
Derrida, Jacques, 18
Devlin, Patrick, 86
Dilbert, 15, 84
Dilemmas, 59, 98, 99
Duke University, 112
Dum-dum bullets, 108
Duncan, Daniel, 67

Earthquakes, 139
Ecole Polytechnique, 139
Einstein, Albert, 19
Engineering failures, 44
Engineering triumphs, 43
Enron, 134
Environmental racism, 121
Ethical egoism, 20, 21
Ethical theories, 19
Ethicists, 32, 52, 59
Euthanasia, 33
Existentialist ethics, 21
Experts, trusting of, 71
External whistle-blowing, 62, 69
Exxon, 95
Ezsorsky, Gertrude, 116

Factors of safety, 41
Famous engineers, 80
Farrington, E. F., 44
Finite element analysis, 39
Firestone, 138
Florman, Samuel, 104, 105
Foreign Corrupt Practices Act, 95
Foreign engineering work, 97
Ford, 137
France, 139
Freedman, Monroe, 35
Friedman, Milton, 58
Fundamentals of Engineering exam, 139

Genovese, Kitty, 7
George Washington Bridge, 28
Genital mutilation, 100
Germany, 70, 106
Gert, Bernard, 8
Gifts, 61, 85
   to universities, 112
Gillum, Jack, 67
Glazer, Myron, 63
Good Samaritan, 7, 124
Graham, Loren, 29
Grant, Ulysses, 80
Green engineering, 17
Gulf Oil, 95
Gulf War, 108, 117
Guns 'N' Roses, 105

Hancock, Herbie, 80
Harassment, 129

Hardin, Garrett, 7
Hayek, Frederick, 124
Hermitage, The, 43
Hildebrand, Wilhelm, 44
Hippocratic oath, 33
Hitchcock, Alfred, 80
HIV, 72
Hobbes, Thomas, 5, 27
Holland, 34, 95
Hoover, Herbert, 80
Howard University, 61
Human rights, 100, 123
Hume, David, 22
Hurricanes, 49
Hussein, Saddam, 125

Iacoca, Lee, 80
IBM, 112
IEEE, 13, 115
Indonesia, 124, 128
Inherent worth, 58
Internal Revenue Service, 35
Internet, 6
Iran, 124
Iraq, 125
Islam, 7
Israel, 124

James, Gene, 62
Jefferson, Thomas, 123
Jokes about engineers, 36
Journalists, 151
Justifiable whistle-blowing, 62

Kalishnikov, Mikhail, 108
Kansas City Hyatt Regency, 66
Kant, Immanuel, 18, 20, 83
Kevlar vests, 108
Kevorkian, Jack, 33
Khomeini, Ayatollah, 125
Kipnis, Kenneth, 51
Koch, Bill, 80
Kohlberg, Lawrence, 73
Koran, 6
Kuwait, 108

Landfills, 121
Landry, Tom, 80
Lawyers, 19, 35, 36
Legal ethics, 35
Lehigh University, 13
LeMessurier, William, 48
Leonardo da Vinci, 28
Leviticus, 4
License, professional engineering, 139
Life, value of, 59
Lightning, 137
Live loads, 40
Locke, John, 20, 27, 123
Lockheed Aircraft Corporation, 95
London, 80
Loyalty, 133

Lund, Robert, 46
Lying, 2, 83

Malaysia, 124
Mahathir, Muhammed, 125
Managers, 15, 46
Mann, Kenneth, 35
Manners, 127, 128
Martin, C. C., 44
Mason, Jerald, 46
Massachusetts Institute of Technology, 112
McCuen, Richard, 73
McCullough, David, 44
McNulty, George, 44
Media, 151
Medical ethics, 33
Mill, John Stuart, 4, 18
Moore, G. E., 18
Moral development, 73
Morality, source of, 4
Moral rules, 8
Morton Thiokol, 46
Music, stealing of, 6
Muystens, James, 60

Named whistle-blowing, 62
Nationalism, 20
Natural rights, 20
NASA, 46, 47
Networking, 81
Netanyahu, Benjamin, 125
Netherlands, the, 34, 95
New York, 48, 49
News media, 151
New Zealand, 60, 79, 99, 127, 128
NIEE, 13, 143
Nielsen, Arthur, 80
Nigeria, 124
Nike, 112
North Carolina State University, 57
North Carolina, University of, 112, 141
Norway, 128
Nozick, Robert, 124
Nurses, 60

Obligatory whistle-blowing, 62
Off-the-record interviews, 152
On-the-record interviews, 152
Oregon, 34
Original position, 21

Paine, Thomas, 123
Passive euthanasia, 33
PCBs, 45
Pepsi-Cola, 112
Peter the Great, 43
Petroski, Henry, 36
Pharmacists, 11
Physicians, 33, 72, 86
Physician assisted suicide, 33
Pirated music, 6
Plato, 18

Pinto, 137
Politicians, 125
Polychlorinated biphenyls, 45
Prima facie moral rules, 3
Prisoner dilemma, 133
Pritchard, Michael, 147
Professional,
    defined, 11
    engineering examination, 139
    engineering organizations, 12
    engineering seal, 41
    origins of, 11
    registration, 133
    respect, 150
Promises, 3
Prudence, 4
Prufer, Kurt, 105
Pruitt Heights, 45

Quinlan, Karen, 34

Rand, Ayn, 124
Rawls, John, 21
Reagan, Ronald, 46
Regan, Tom, 57
Registration, professional, 139
Reinforcing bars, 40
Rensselaer Polytechnic Institute, 13
Requirements in a code of ethics, 15
Respect, of the profession, 150
Reverence for life, 111
Rights, 20
Roebling, Emily, 44
Roebling, John, 44
Roebling, Washington, 44
Rome, 43
Rowan University, 112
Rules of Professional Conduct, 35
Russia, 13, 108

Safety, 135
Sandia Labs, 108
Sartre, Jean-Paul, 18
Scandinavia, 129
Schweitzer, Albert, 111
Seal, professional engineering, 41
Self-laudatory language, 90
Seneca, 124
Shaw, William, 96
Sherman Antitrust Act, 92
Silkwood, Karen, 63
Singer, Peter, 7, 56
Situation ethics, 21
Slavery, 100
Spice Girls, 105
Social contract, 5, 27
South Africa, 108, 124
Soviet Union, 29
St. Petersburg, 43
Stanford University, 141
Stevens Institute of Technology, 134
Stevenson, Adlai, 79

Strangers, obligations to, 7
Stubbins, Hugh, 48
Suharto, 125
Sununu, John, 80
Superogatory behavior, 7
Sustainability, 17
Sustainable development, 17
Sydney Opera House, 38
Sympathy, 22

Tarasoff, Tatiana, 99
Tax lawyers, 35
Taylor, Paul, 57
Tchaikovsky, Peter, 106
Teaching assistant, 84
Ten Commandments, 4, 20, 21
Tenure, 78
Term papers, 6
Terrorists, 28
Thames River, 80
Thailand, 71, 83
Thomas Aquinas, 18

United Brands, 95
United Kingdom, 17
United States Department of Justice, 92
United States Military Academy, 13, 80
Universalizability, 20

Use of animals, 57
Utilitarianism, 20, 22
USSR, 29

Veil of ignorance, 21
Value of life, 59
Vaughan, Diane, 47
Vegetarianism, 110
Vermont, 7
Virginia, University of, 85, 141
Vikings, 147
Virtue ethics 21

Washington, George, 13, 80
Wasserstrom, Richard, 19
Weapons, 107
Werhane, Patricia, 63
West, Frank, 95
West Point (United States Military Academy), 13, 80
When-in-Rome clause, 97, 127
Whistle-blowing, 12, 62
Wigand, Goeff, 63
Wisconsin, University of, 78
Working together, 37, 53
World Bank, 17
WorldCom, 134
World Commission on Environment and Development, 17
World Trade Center, 28